DEVELOPMENT COALITIONS IN WORKING LIFE

Dialogues on Work and Innovation

The book series *Dialogues on Work and Innovation* presents empirically based studies as well as theoretical discussions on the practice of organizational renewal.
Its publications reflect the increasingly urgent need for the development of new forms of work organization. In today's interdependent world, workplace reform and organizational effectiveness are no longer solely the concern of individual organizations; the local and the global have become closely interconnected.
Dialogues on Work and Innovation mirrors the fact that enterprise development and societal development cannot be kept separate. Furthermore, the Series focuses on the dialogue between theory and practice, and thus on the mutuality of knowledge and action, of research and development. The *Dialogues* stress the critical significance of joint reflexivity in action-oriented research and the necessity for participatory processes in organizational change.

Editor
Hans van Beinum
Centre for Working Life Research and Development
Halmstad University
Box 823, S-301 18 Halmstad
Sweden
Hans.van_Beinum@cau.hh.se

Volume 6

Björn Gustavsen, Tom Colbjørnsen and Øyvind Pålshaugen (eds)

Development Coalitions in Working Life
The 'Enterprise Development 2000' Program in Norway

Development Coalitions in Working Life

The 'Enterprise Development 2000' Program in Norway

Edited by

BJÖRN GUSTAVSEN
TOM COLBJØRNSEN
ØYVIND PÅLSHAUGEN

JOHN BENJAMINS PUBLISHING COMPANY
AMSTERDAM/PHILADELPHIA

The paper used in this publication meets the minimum requirements of American National Standard for Information Sciences — Permanence of Paper for Printed Library Materials, ANSI Z39.48–1984.

Library of Congress Cataloging-in-Publication Data

Development Coalitions in working life : the "enterprise development 2000" program in Norway / edited by Björn Gustavsen, Tom Colbjørnsen, Øyvind Pålshaugen.

p. cm. -- (Dialogues on work and innovation, ISSN 1384-6671 ; v. 6)

Includes bibliographical references.

1. Research, Industrial--Economic asects--Norway. 2 Technological innovations--Norway. 3. Organizational chage--Norway. I. Gustavsen, Björn. 1938- . II. Colbjørnsen, Tom. III. Pålshaugen, Øyvind. IV. Series.

HC370.R4R46 1997

338'.064'09481--DC21 97-45677

ISBN 90 272 1776 9 (Eur.) / 1 55619 830 2 (US) (Pb: alk. paper) CIP

John Benjamins Publishing Co. · P.O.Box 75577 · 1070 AN AMSTERDAM · The Netherlands

John Benjamins North America · P.O.Box 27519 · Philadelphia PA 19118–0519 · USA

Shipping Manifest

Phone: 0000000000

Sales Order No.: 774610379
AbeBooks Purchase Order No.: 738117709
Processed: September 11, 2026 **Delivery Time:** 10 - 14 business days

Item	Author	Title	Bookseller Inventory No.
1	Gustavsen, Bjorn; Colbjornsen, Tom; Palshaugen, Oy	Development Coalitions in Working Life: The Enterp	047527
	Book Description: 153pps. Card wrapper is clean and without wear. Binding square and tight, not creased. Octavo. Volume six from the series, Dialogues on Work and Innovation, Hans van Beinum, Gen. Ed.		

Please retain this packing list for your records. If you have any questions about this order, please contact the bookseller using the Ask a Bookseller a Question link available in your AbeBooks account.
**Please Note: AbeBooks has collected payment for this order.

Thank you for shopping with AbeBooks.com.

AbeBooks

Acknowledgements

This book emanates out of a Research and Development program jointly supported by the Norwegian Research Council, the Confederation of Business and Industry and the Federation of Trade Unions. We would like to express our appreciation for the willingness of these institutions to invest in a workplace development programm focussing on work organistion, participation and workplace development. Our acknowledgements also go to the board of the program for help and support in building the complex and difficult bridges that such a program demands, between research, workplaces and the labour market parties. Several coleagues have helped evaluate and improve on the contributions to the book: we will, in particular, mention John Shotter, University of New Hampshire; Werner Fricke, Friedrich Ebert Stiftung; Philip Cooke, University of Wales; Frieder Naschold, The Science Center, Berlin, and the editor of the book series: Hans van Beinum.

Oslo, January 1998

Bjørn Gustavsen
Tom Colbjørnsen
Øyvind Pålshaugen

Contents

Preface

Norwegian enterprises confront, like enterprises in all societies, the challenge associated with global competition such as the ability to continuously differentiate the products as well as continuously improve on performance within fields like production time, logistics, quality and much more. In this process a broad mobilization of resources need to take place within each organization as well as within the surrounding society.

Within the organization, the motivation and commitment of all the employees is crucial. On this, there is little disagreement today. To actually create this commitment and motivation is, however, not so easy: there are different views on how this should be done.

And even when people actually are committed to the goals of the enterprise and eager to do their best, there is a need for a rational organization that allows them to actually give their best. A faulty organization can lead even the most motivated into blind alleys and unproductive efforts. At this point the issue of work organization enters the picture — another topic where there emerges a general agreement on its major significance, but not a similar agreement on what to actually do: what structures will organize people in such a way as to ensure the maximum productive and innovative outcome from the efforts they make?

In this kind of situation it is natural for the enterprises to turn to whatever sources of knowledge may exist in society. Among these, research is a major one. Departing from the problems sketched aboveit is research on work, organization and leadership that emerges as most appropriate. But how can the enterprise utilize this kind of research?

In taking a closer look at this research it is seen that this is not so easy, either. It tends, for example, to be organized in different, and often competing, schools of thought. How, then, to decide which one is most appropriate? Even if the knowledge looks adequate it can, furthermore, be difficult to give it a form that makes it accessible to workplace actors. Workplace actors need to think in terms of processes of continuous improvement rather than in terms of one step transformations. To use research, then, means to establish a relationship that can prevail over time and provide a basis for discussions. Are such discursive relationships possible? If research moves into a number of relationships to different local actors with different demands: how can it maintain a role as responsible for knowledge and enlightenment on a more general level?

If the utilization of research to help enterprises deal with their problems is to be of any importance for the development of work, organization and industrial relations on the level of society it must imply more than a few scattered contacts. The contacts, or relationships, between research and enterprises must have a certain magnitude — there must be a number of researchers working with a number of enterprises. This, however, calls for institutional solutions to the problems emerging in creating these relationships. There must, within the academic community, be an acceptance of the kind of role into which research has to enter to be able to help the enterprises. The enterprises must, on their side, accept that there are basic differences between using research versus using consultants. Research has a long term development of publicly available knowledge as its main goal and needs, in this context, to get a number of principles accepted as part of its relationships with the enterprises, such as the possibilities for utilizing experience from joint development work as a basis for research and associated activities, such as publishing results and analyses.

These are examples only of issues and problems emerging when bringing research in as a resource into the processes of continuous improvement which all enterprises now have to initiate and sustain. When the issues are many and the problems difficult there emerges a need for organizing the agenda-setting and problem-solving process itself in a fruitful way. There is a need to create a context where enterprises and research can meet and jointly discuss how to relate to each other; there is a need for arenas for research to discuss internally and there is a need to link all these activities to actors and institutions with responsibility for the overall patterns of co-operation and industrial relations in society.

"Enterprise Development 2000" aims at covering this ground. As an initiative from The Norwegian Research Council, the program emanates out of the long established co-operation between the main labor market parties in Norway. It involves a substantial number of enterprises and researchers. Organized in "modules", groups of researchers are linked to groups of enterprises — together forming "development coalitions" of which the program encompasses seven, each with its own characteristics. Each such coalition can be seen as local. However, the program also aims at transcending the local to create more general perspectives. This is, however, not to be done through enforcing one single understanding on all the experiences to which the program gives rise, but through promoting discussions: between the researchers internally and between the researchers and the labor market parties.

In sum, rather than reflecting a fixed view on how to give research support to workplace development processes, the program can be seen as a learning expedition. As such it encompasses a lot of participants and is in this sense a joint venture of some magnitude. It is this size — or mass — that gives the program its potential, for instance for exploring a substantial number of issues within the field of work organization, leadership, co-operation and inter-enterprise relationships, and for circulating this "mass" of perspectives into processes of evolution in working life. This kind of approach, where discussions are emphasized, along with the possibilities for all concerned to participate, and with research in a supportive rather than a dictatorial role, is seen as consistent not only with the type of knowledge we deal with in a program like ED 2000 but also with the democratic traditions of Norwegian working life.

Ottar Henriksen
Executive Vice President
Hydro-Raufoss Automotive
Chairman of Enterprise Development 2000

Chapter 1

Enterprise Development 2000

A Program for Distributive and Interactive Learning

Bjørn Gustavsen

Introduction

This book has its origin in a program devoted to the improvement of work and organization. The Enterprise Development 2000 program, as it is called, is located in Norway and is an extension of the co-operation between employers and unions which has been a salient feature of Norwegian working life since the immediate post World War II period. ED 2000 is not only a bipartite effort, however, since a research element is also involved, having a real role to play, so there are actually three parties involved. In this sense, the program links to the series of research based development efforts that have taken place in Norwegian working life since the 1960s (for a review see Gustavsen, 1996).

Formally housed within the Norwegian Research Council, the program was initiated by the Norwegian Confederation of Business and Industry (NHO) and the Federation of Trades Unions (LO) as an extension of their joint activities in support of workplace and enterprise development. Currently, the program consists of seven research groups — several of which are actually composite groups, with members drawn from different institutions. Formal agreements on joint development efforts have been signed with about 60 enterprises. The majority of these are members of networks or other systems for inter-enterprise co-operation, where the whole network is accessible for co-operation with the program if resources and other circumstances favor such an expansion. Each group has in principle to place five research-years at the disposal of the program. Since these tend to be spread over more than five researchers, the number of researchers who have ED 2000 as a main — although not exclusive — activity is well over 30. In addition there are more than 15 doctoral students linked to the program. This is in line with one of the major intentions of the program, which is to build up not only competence but also organization within this field in such a way that, in the future, it will be possible for any enterprise within the LO-NHO area to become a member of a research-supported development coalition with other enterprises, wherever the enterprise may be located in physical and social space.

Each combination of researchers and enterprises is called a 'module' for reasons to which we shall return later.

Each 'module' can also be called a 'development coalition' — a concept that can be applied to the program as a whole, too. We shall also devote closer attention to the background to this concept later.

A point of departure of the program is that it pays half the salary of the researchers involved — many of the doctoral students are on special budgets — plus ordinary overheads. This limiting of support to half the salary is not so much motivated by a need to save money as by a need to have other interested parties involved on the financial side. A program like ED 2000 cannot fulfil its mission unless the activities generated by the program can be carried on beyond the time frame of the program itself (Naschold, 1993). The participating enterprises pay their own costs. The principle underlying this is that the enterprises are supposed to participate because they need a certain kind of development — a need that exists independently of the program. Furthermore, with the kind of total, or all-encompassing, processes of development characterizing the modern scene there is no way in which a public research council could undertake the financing of 'project costs' anyway. A number of the developments with which ED 2000 is involved could, alone, easily swallow the whole budget of the Norwegian Research Council.

The program has its own board, made up of representatives from the parties in the labor market, business, and research, with a secretariat to deal with operational issues, having a parallel composition. The program is scheduled to run from 1995 upto and including the year 2000. It has an international benchmarking team, with members from Germany, France, Sweden and the UK, in addition to Norway.

Background

When the issue of work organization originally emerged as an important topic, the focus was on what pattern to strive for — what type of work organization would be most appropriate from, say, a combination of participation and productivity perspectives. This focus has gradually changed: as the number of good examples, interesting solutions and, for that matter, theories, has grown, the question has increasingly become how to get the views applied. If one company has developed a new and more productive pattern of group organization, say, how can one get others to apply the same pattern?

According to a classical notion of enlightenment, a good description and documentation of the innovation — preferably a 'scientific' one — should in principle be sufficient. When 'the best solution' is known there is a moral obligation to apply it. However, without recapitulating the long and complex history of work life change and reform, we know that this simple, linear diffusion from innovative cases based on the binding force of 'the best solution' did not automatically come into existence.

This gave rise to a new issue, generally called 'commitment to change": for new solutions, patterns and innovations to be diffused and form the basis for change in new and wider circles of workplaces there must be an interest in, and a commitment to, undertaking this kind of change. How, then, do we create commitment to change? What *is* commitment to change? Do we talk about psychological states in individuals, tendencies in groups and other collectivities, or institutional characteristics of organizations? The answer is important since it determines the approach to choose when seeking to generate commitment. Furthermore: how do we distinguish between commitment to change and the actual performance of change? This question arises because it is reasonable to assume that the one who is committed to change is also involved in change. It is hard to imagine a strong commitment together with a complete passivity. If this is so, many of those committed to change will already have their own experiences, ideas, practices and processes. How do we relate these to the ideas that are to be diffused? Do we argue that the ideas we can offer are better than those they pursue themselves? If so, we open the game by criticizing them — perhaps not the best approach. But if we do not chose this dubious opening gambit, what do we do then? It seems that the only answer is that there is a need to make a *serious, open, reflective comparison* between what they already do and what the new ideas are offering.

The issue of achieving 'mass', or scope, in workplace change involves at least three main issues: ideas, diffusion channels, and commitment. These issues seem, furthermore, to be linked: if what happens in one workplace is to achieve significance in other workplaces there is a need for channels through which information can move, and such channels, in turn, depend on an interest in, and a commitment to, change. If, however, it is assumed that a commitment to change is, at least in many cases, linked to actual involvement in change, we are confronted with a circle. Those most interested in what happens in one organization under development are those who are themselves located in other organizations that are also under development. Clearly, there will be modifications and exceptions to this, but this interdependence between commitment and doing seems to be of such importance that, unless our thinking about change captures this specific configuration, we will probably fail in general.

There is also a fourth element of major significance concerning how to approach this bundle of relationships, namely that the belief in patterns, or solutions, applicable throughout working life as a whole — universal reason — has not stood up too well to the critical examination to which it has been subjected during the last two decades (Gustavsen, 1996). The dominant view today is that there may be one best pattern if a specific context is given, or that general theory and ideas have something to offer in specific processes of change, but only as elements, or parts of, the 'raw materials' out of which local solutions can be constructed, not as complete packages. The idea of a universal reason, more or less directly applicable everywhere, is no longer feasible. Consequently, there is a need to approach the case-diffusion-commitment bundle in such a way that it allows for a number of cases/ideas/experiences, which can promote a range of different patterns, to move around.

A continuous focus on the single case, without considering the other elements in the bundle, seems to be of limited fruitfulness. Likewise, to focus on diffusion and commitment separately would not be fruitful either. In general, however, there is far more available in terms of case studies of single-organization change than there is concerning diffusion perspectives or those pertaining to commitment. The major deficits are on the last two points.

The links between the three basic elements — under contingency conditions — can be forged in different ways. In Scandinavia there is a tradition of creating programs. A program involves bringing a number of enterprises into a joint arena where they can create new solutions and exchange them with each other, often with the help of some external resources such as researchers or consultants who can help interpret experiences, establish diffusion channels, and assist in exploring ways of using experience together with the workers and managers in the participating enterprises.

Since such programs will generally attract participants with an interest in — and a commitment to — change, together with some pre-existing experience, they will not contain representative samples of all enterprises in society. This, however, is no major shortcoming. The idea of representativeness, in a statistical sense, has little importance in strategy for change. There are many other types of relationships between enterprises than the extent to which they resemble each other that become of significance in a process leading to a critical mass.

Within the framework of programs it is relatively easy to create and sustain diffusion channels, or arenas for joint discourse, not to go one step further and say that a program can be shaped to actually *be* a set of such arenas.

Third, a program includes, by definition, a number of cases that can form platforms for exchange of experience and mutual learning. One objection to relying on the experience which just happens to be present within such a program is that 'the best practice' may not be present within the participating population — in fact it is highly unlikely that it is if we view best practices in a global perspective. Building a regional network of, say, 50 enterprises somewhere in

Scandinavia, or for that matter anywhere else, will, it can be argued, constitute far too narrow a basis from which to learn. This objection confuses two issues: on the one hand what ideas to apply, on the other how to create processes of change. While a network of 50 enterprises means that it is just these 50 enterprises that work together in this particular network, the ideas that can be brought into the process are in no way restricted to those generated within the network. Import from outside can take place freely — in a network as well as when the unit of change is the single enterprise. The point about a larger network is that it can actually generally import more ideas and be more open to the global than is possible for a single company. To adopt and use an idea in a process of development is not something that is done by reading about it in the *Harvard Business Review* in the morning and having it implemented in the afternoon. Instead, it is a long process of search, understanding and stepwise development of action. And since the world is full of potentially relevant ideas the limit of what each enterprise can explore on its own compared to the total potential is soon reached. In a network a broader range can be brought into the process and the ideas can be compared with each other, step by step, as they emerge, when transformed into the particular local context constituted by the network.

It is for reasons like this that not only specific development programs but inter-enterprise co-operation in clusters, networks, regions and the like are mushrooming today. Some of them are large, like the much discussed Emilio-Romagna region in Northern Italy with about 4 million inhabitants; at the other end we find clusters of, say, a dozen or so small to medium sized enterprises with maybe a few hundred people involved. While there are great differences between these constellations they have something in common: they represent a 'flocking' of enterprises to arenas where they can meet some of their challenges together. Some of the configurations are so limited in mass that they hardly constitute challenges beyond those that appear within single organizations. Others have more 'mass', and genuinely constitute, or approach, a 'meso-level' in society.

As a point of departure, such meso-level formations emerge as answers to the critical problem of how to create scope, or mass, in the changes. Thus far the answer to the question of how to diffuse new forms of work organization is: create a network. Unfortunately, in this field there is no ultimate answer, in the sense that it indicates a solution which is itself free of problems. The emergence of formations that in themselves constitute a reasonable mass actually lead into a new set, or generation, of problems:

First, 'political' issues emerge, such as the relationship with collective agreements and the role of the unions and employer organizations and also of the state and the order it has chosen to create in working life. While experimental changes in small workplaces can generally be exempted from this kind of setting, larger systems, even if they are on the move, cannot be exempted from the ordinary regulations, in particular not when the number of such coalitions is growing. Consequently, a need emerges to grapple with this kind of issue, in the form of a *political function that can work on line in relation to the network* at the same time as it is able to mediate between the network and society level institutional frame conditions. When change transcends the level of the single case — however much of a 'beacon' this case may be — and moves towards 'critical mass' the question of the interplay between the change processes and the surrounding society grows in importance. In a study of the Swedish Working Life Fund the idea of 'the innovation supportive society' as something to move towards was applied in this context (Gustavsen *et al.*, 1996).

Second, a new generation of problems emerges, related to the knowledge generation process and to such resources as research. If we presume that one of the main roles of research is to help understand and give expression to processes and structures in a development context, at the same time as such a context shows a plurality of processes and structures, research must be able to work with such a plurality. And what does that mean? It does not mean an automatic attempt to bring them all under some kind of common denominator. This may be the outcome of a process of testing and comparison, but the point of departure has to be different. *It has to place its main focus on grasping, understanding and elaborating each alternative that is present within, or accessible to, the coalition, on its own conditions and premisses.* And in approaching this task we have to depart from a radical contingency theory where we cannot presume that the same concepts and variables are to be used (and differences reflected in the form of different values on the same variables). Rather, we must depart form the assumption that 'in the beginning' *different conceptual schemes are needed to capture each of the processes and its structural outcomes.* An obvious consequence of this is that research cannot be committed to one single conceptual or theoretical framework. At the same time, however, there is the same demand here, as elsewhere, for enthusiasm and involvement from the side of research, for believing in the value of the phenomenon one works with, and for being willing to help promote it. How do we solve this dilemma? One way of doing it is to utilize different researchers or research groups within one and the same program, where each researcher/research group can help to elaborate one specific alternative within the broader coalition. Then, however, no group can play the superordinate role of bringing the alternatives together for comparison. How do we achieve this? In principle, the answer is simple: through a *discourse*. This discourse must actually be an integral part of the more general discourse about all the experiences and interpretations made within the network. In actual practice, the answer is quite complex. In social research, generations of researchers have worked with the aim of forcing different phenomena to constitute elements within one and the same theory — now we have to accept a situation in which the researchers are encouraged to elaborate different theories at the same time as they are to join in a discourse where the aim is to compare, to make choices and maybe to discard some of the alternatives as less fruitful. How is this to be done?

The problem of *the on-line political function* and the problem of *the role of research within a development context based on discourse and experiential as well as theoretical pluralism* are the two core dimensions in the 'Enterprise Development 2000' program. These are the two issues on which the program places special emphasis. In addition, it is, of course, expected to provide a new generation of experiences from — and perspectives on — a number of more classical topics, such as the use of research in a development context, linking back to the notion of action research; what relationships ensure the most fruitful exchange between research and enterprises; how to understand and construct inter-enterprise co-operation with a developmental purpose; the potential and future of the Scandinavian type co-operation between the labor market parties, and more (we will return to some of these issues later in this chapter; more will emerge from later chapters).

How important are the two issues that emerge as the core issues of the program? Are they two issues among a range of issues where all have equal importance, or do they carry more weight?

For anyone familiar with the contemporary discussions within fields ranging from the regionalization of economies, such as the European nation states, to such issues as the role of theory and understanding in the social field, it is easily seen that the terrain to be covered by ED 2000 is the core terrain at this time. At least in Europe we see the paradox that at the same time as global competition and interdependencies are growing, so does the need for 'home base' network support and alliances. The global and the local are, in terms of social relations, not a zero sum situation but one of mutual reinforcement. Consequently, the need to give a political recognition of and umbrella to this development is already substantial, and growing every day.

In departing from a pluralist view of theory and best solutions one actually places oneself in what is no longer an oppositional school of thought but rather the mainstream. In the age of post modernism, post structuralism and deconstructivism the potential of general theory and universal reason has been challenged from a number of angles. The more 'grand theory' is undermined the more, however, the question of *what do we then actually do* forces itself on us. If there is no single best theory we need pluralism. But there is often a need to bring a large number of actors together around some kind of common cause or understanding, not least, by the way, in working life with its ongoing globalization of production and markets. How do we achieve this? How do we move from the single case to the larger system if we cannot do it through hierarchically structured general theory that in some way or another is to give a 'top down' validity to all individual experiences? As indicated above, the procedure that offers itself is 'the open discourse", where options are identified and positioned in alternative figure—ground relationships to each other so that choices can be made and further development performed. This is reasonably easy in theory. But how is one to do it in practice, given the conditions indicated above? Even small networks may contain a great deal of experience and it can need a substantial effort to draw it out. The comparative positioning of alternatives in relation to each other is no simple process since the main outcome of such a comparison is rarely a simple choice between given alternatives but rather the emergence of a new generation of alternative courses of action. If a network or other coalition is of limited size it will have to link up with other coalitions, increasing the complexity. Out of any kind of coalition of specific enterprises in a specific context there will be threads linking it to the global, creating a continuous traffic across the boundaries and a corresponding need to lead this traffic into the internal channels of the network. Of course, such a program as ED 2000 cannot demonstrate all aspects of how to do this and, in addition, do it on a globally significant scale. In this perspective it is a small program in a small country. But the work with this kind of process on the practical level has to start somewhere, however, and as programs go ED 2000 has a certain mass. It includes a substantial fraction of the major industrial-economic actors in Norway as well as a number of medium sized but highly competitive enterprises, which provides for a reasonable selection of participants. It is possible, through such a program, to help identify, analyze and gain experience with, some of the steps, some of the building blocks, that need to be considered for politically significant, discursively oriented, development coalitions to grow.

Program phases

A program of this kind can, somewhat mechanistically, be seen as consisting of certain phases:

First, a phase of constitution, in which a structure must be created with the following potential:

- access to a range of relevant experiences that can function as 'raw material';
- the possibility of elaborating and playing these experiences into new processes of development;
- initially on their own premisses;
- but with a view to entering broader streams of efforts were comparisons and other forms of interplay are possible;
- the on-line use of research;

- on-line links to political-institutional actors in such a way that the processes of the coalition can have a formative impact on industrial relations systems, agreements and public regulations and policies.

Second, a phase in which the various boxes are filled with actors, such as enterprises, researchers, and representatives of institutional forces.

Third, a start up phase in which the wheels are set in motion and the various relationships turned into living activities. Critical issues here are:

- to establish working relationships between enterprises;
- between enterprises and research;
- to a degree sufficient to ensure a reasonable 'mass' on program level as well as a fruitful range and mix of experiential-theoretical alternatives;
- a decision making process at the program level that implies a fruitful interplay between enterprise/network activities on the one hand and political/institutional considerations on the other.

Fourth, a phase of continuous work but also of linking the different combinations of experiences, interpretations and actions to each other in joint arenas.

Fifth, a concluding phase of identifying what has been learned from the program with a view to institutionalizing this learning as well as consolidating patterns for use as platforms in continued processes of change.

As 'Enterprise Development 2000' emerges in this book, the constitutive phase is covered as well as the start up phase and some elements of the phase of ongoing work. It is possible to follow the different research groups on their way into the enterprises to see how the relationships are evolving. Some elements of the political processes surrounding the program are covered, in the sense that what can be demonstrated is the way in which actors, topics and discourses need to be shaped for a fruitful political discourse to potentially emerge.

It is, in other words, in many ways a 'halfway-house' report. Why not wait until the program is concluded, among other things to be able to go more deeply into the core problem of creating discourse between different theoretical-experiential positions to reach some kind of common ground? The reason is that the program not only aims at creating discourses; it also aims at linking to discourses, among them ongoing ones within the philosophical and social science community around the role of social theory, social understanding, and social knowledge. Perspectives like those upon which ED 2000 is built are, as mentioned above, not unique to this program. Its claims to uniqueness are primarily found in its effort to give *certain arguments a practical expression, to see how they function in real life*. To exemplify some of the links, it is possible to mention views like those expressed by Giddens (1990) on social theory as a plurality of interpretations that weave themselves in and out of practical situations; those expressed by Toulmin

(1990) on working together without letting the work be mediated by one single superordinate theory, those expressed by Shotter (1993) on the need to use social theory as a help to see things, create figure-ground relationships, and the like, rather than conferring eternal truths. To this can be added a number of points emerging out of the debate on — and experiences from — action research, such as the need to work together with those concerned rather than from a distance, stressed by many contributors such as Whyte (1991), Greenwood & Santos (1992), and Fals Borda (1987); the need to see research as a collaborative inquiry done together with those concerned as argued by, for example, Reason (1993); the idea of constitutive dialogue as the core element in the relationships between research and those concerned (Isaacs, 1993; Gustavsen, 1992) and the need to develop a social science that can function 'on line' and in 'real time' in relation to the problems with which it deals, as argued by Argyris and Schön (1978). It is our intention to position ED 2000 within this broader debate, to become one of its sources of experience. What meaning can be assigned to the experiences that emerge out of ED 2000 in terms of contributions to the broader debates is, however, an issue that also needs to be approached discursively. These meanings will not emerge as a limited list of fixed conclusions on the final day of the program. This publication, then, is primarily intended to introduce ED 2000 as a reference point in ongoing discussions. It aims at giving those interested sufficient familiarity with the program to make it possible to relate to it and link it to their own experiences.

The interplay between workplace processes and institutional context

If any potential influence is to be exerted on conditions in the political-institutional frame, it is necessary for a program to perform a linking function between what happens at workplace level and what happens with the shaping of the relevant institutions in society. To be able to perform this function, Enterprise Development 2000 is linked to an agreement on development between the LO and the NHO. Entered into in 1983 and revised in 1990, the agreement expresses the intention of the central federations to promote and support local development efforts that can help enterprises in their processes of value generation (Gustavsen, 1993). Such support presupposes that the development is defined as a joint one between management and labor and that certain procedures are applied in this context. By labor we do not just mean participation through union representatives. The agreement emphasizes the demand for direct — and hence broad — participation. 'Broad participation' in local processes of development has increasingly come to be seen as a major and necessary part of the idea of industrial democracy. Consequently, the program links to the tradition of efforts to democratize working life that have been going on in Norwegian working life since the mid-1940s. At the same time, however, it aims at a full recognition of the ongoing internationalization and the need for the enterprises to be competitive in a world market. A primary function of the agreement was to open a new 'space' in the industrial relations system: an arena in which certain issues could be taken up, discussed and decided on within an institutional framework.

Through the link with the agreement on development and, in turn, to the whole system of agreements and established practices of co-operation and conflict between the labor market parties, the program actually operates within a field that is rich in rules and traditions, all of which are expected to be known and used by all actors in the program. It is this embeddedness, on the other hand, that makes it possible for the program to have an impact on its own institutional-political context.

The researchers must not only demonstrate an ability to come to terms with local management. In ED 2000 there is a need to come to terms with the local unions as well, but not only this: given the demand for broad participation there is actually a need to come to terms with all the employees. Furthermore, the relationships with the social partners have to be created not only on the local level but also on the national level, and generally on regional levels as well. Research, in other words, is involved in a network of relationships with the actors in the industrial relations system.

But what is the purpose of creating a specific program to get research to be steered by aims and considerations of this kind? Could not these be imposed at the project level and, insofar as what happens in the projects is to be assigned any significance for the agreement as such, why should these experiences not be fed back directly from the projects to the bodies in charge of the agreement? The answer to this is simple but important: no constraints nor aims can ever be expressed in an absolutely unequivocal way and, in addition, in a way that remains immutable over time. A need such as that for 'broad participation' is not an arithmetical rule, to be applied mechanically — rather, it demands an interpretation of each situation, in addition to in itself being subject to evolution. For these interpretations to be performed and for the evolution to be the result of a process where reflection plays a major role there is a need for an *ongoing dialogue* between research and the labor market parties. The link between workplace development processes on the one hand and institutional conditions on the other is not a mechanical transmission but is forged by acts of construction. The program is to constitute the arena for such a dialogue in which the labor market parties accept that they must continuously consider the points, arguments and experiences of research, and vice versa.

Since the research groups depart from different theoretical perspectives, the labor market parties can, in their dialogues with research, easily be exposed to a number of *different* views and perspectives promoted by different groups of researchers. If it is to make an impact, does not research have to present a unified view? Not necessarily — if we look at the history of the labor market parties the most striking feature is that they appeared with very different 'theories' concerning their *own* mission in society and what needs to be done to instantiate them. In fact, within the landscape of classical political dividing lines there are few groups that stand further apart than the employers and the unions. Notwithstanding this, efforts like the program described in this book have, together with a lot of other efforts, emerged as joint ones. The ability to launch joint efforts does not seek its reason in the abolition of all 'historical differences'. They are present today, to a greater or lesser degree, just as they have always been. The joint efforts emerge out of an ability to compromise — to reach agreements exactly *in spite* of different points of departure. The labor market parties are not only able to operate without having to be guided by the same 'universal reason'; they are actually particularly well prepared to move ahead just when this universal reason is *not* present.

Bringing different views together

Whereas the design of ED 2000 is based on theoretical and methodological pluralism, it does not *prohibit* general perspectives from emerging — be they perspectives that can be called 'theory' or other types. In fact, the program encourages the development of such perspectives. It does, however, do it in a certain way: through a joint research agenda and through the setting up of arenas where the researchers can meet each other to discuss their respective positions on the topics of the joint agenda. In this way 'general theory' can emerge, but only after a process of dialogue through which the researchers agree to join in a common view. *Agreement* is, in other words, the key element rather than one single 'scientific logic' which is supposed to be binding on all.

The agenda on which the program wants all groups to focus — generally in addition to topics of their own choosing — contains the following points:

1. How are viable and fruitful development processes to be created at workplace and enterprise level?
2. How are the development processes/development organization to be linked to each other in a number of enterprises, so that interorganizational processes — networks, dynamic regions and the like — can be formed?
3. How may concepts, frames of reference and international co-operation be developed to facilitate an improved integration between Norwegian enterprises and the front line of international developments within areas like quality, logistics, product development, and the like?

4. What role does the Norwegian/Scandinavian traditions of labor-management co-operation play in strengthening the enterprises in the international competition — if any? Is there a need to restructure the labor relations system and patterns of co-operation and participation?
5. How can business strategies and business ideas be developed that can provide platforms for mobilizing the employees around the goals of the enterprise?
6. The state/the public offers a number of systems in support of enterprise development. How do these systems work — for instance in terms of ability to support long term and coherent strategic processes at enterprise or network level?

It will be seen that these topics, despite their descriptive form, lean towards the process side. The aim is to shift focus towards what to do, how to organize, how to link, how to relate. The topics are those that emerge as most critical to the further expansion of the activities under the agreement on development. There is not necessarily one best answer to each of these questions; the point about a joint research agenda is to pull the researchers into a field where they are brought to consider the same issues and are at least forced into as close a relationship with each other as possible.

The role of research

Given the outline of the role of research within the program emerging above, how should this be characterized? In its link to specific processes of evolution in specific workplaces, the research role resembles that inherent in the notion of 'action research'. The extent to which this word should be applied (or not) is a matter of taste and position; in the contributions to follow some use it and some do not. The involvement in concrete workplace processes with a view to helping improve them corresponds to the original notion of action research (Lewin, 1947). So does the use of 'does it work' as the primary validity criterion. In most social theory the so-called correspondence criterion is applied when assessing the quality of a theory — does it adequately reflect, or mirror, reality? In action research and its offsprings this criterion is not the core one. At this point action research is in line with the general tendencies in most theory within fields like physics and biology, where the main test of a theory lies in seeing to what extent it allows us to perform certain acts with certain consequences in reality. On the other hand, since we no longer work with experiments with a clear delimitation in social space, the 'does it work' criterion is not a simple one to apply. How, for instance, do we answer questions like 'does the historically given pattern of co-operation in the Norwegian labor market work today'? The fact that the validity criterion is not always easy to apply means, however, that its use has to be worked on and further developed, not that we have to revert to a correspondence criterion.

In applying the concept of 'action research' one may easily add the term 'participative', which has become quite common (i.e. Whyte, 1991; Fals Borda, 1987). With its emphasis on problem-driven processes and on the major role played by 'the users' the notion of 'participative' seems also to be appropriate to ED 2000. On the other hand, it depends on what is put into this word. If 'participative' is taken to mean that no autonomous role at all is possible for research and that there are no fields where research can play a specialized role within a broader context, the participative notion does not fully apply to ED 2000. It is a main purpose of the program to create policy platforms and linguistic frameworks that can enable research to actually proceed on its own, i.e. in investigating the issues on the joint research agenda. Even though many of the inquiries, studies, or investigations performed within the program need to be collaborative, not all need to be performed in this way. For this reason the concept applied in the title of the book is held out as perhaps more appropriate: the idea of a *development coalition*. In such a coalition there are a number of actors who develop joint platforms and frameworks but who otherwise function on the basis of complementarity: management manages, union representatives represent, workers take care of shop floor processes, and research researches. When they meet in joint arenas to create common platforms they consider each other's points and arguments, but they are still there in the capacity of their ordinary role. When management and unions agree on something it is not because they have 'taken over' each other's view, or even, of necessity, developed a common view; it is because they have found it reasonable and in their own best interest to support each other in a certain course of action. This is the kind of context within which development research in working life has to identify its role and contributions. If research and those people that are concerned with the activities of research — viz., the actors in working life — can agree on the ways in which a certain set of words are to be used, research can fill more ordinary research functions within this framework. If research and the labor market parties can agree on, say, the significance of group work for the idea of industrial democracy, the actual performance of a study to investigate the scope and depth of democratically relevant group developments in a sample of workplaces can be left to research. Within such a framework research can be objective as well as autonomous.

Since a process of research and development in working life has to cover a number of interrelated as well as diffuse topics — productivity, quality, participation, globalization, benchmarking, just to mention some of the contemporary buzzwords — it follows that the agreement-reaching capacity of the program must be substantial. Compared to social research as traditionally defined, this constitutes a reversal: in 'positivism' it is thought that objectivity needs a maximization of the *distance* to those concerned, whereas in ED 2000 it is rather the number and intensity of *joint discussions* that are subject to maximization. The view of 'objectivity' is reversed: unless research can operate within a framework of agreements, no objectivity is possible since research will express its thoughts in a language the meaning of which is not shared by the other actors. The program provides the platforms, or arenas, for the working out of joint linguistic frameworks within which research efforts can occur. It can, however, not commandeer such agreement — it can only hope that a joint research agenda, in combination with a number of arenas, will be sufficient to generate new perspectives as well as reasonably broad agreements on the validity of the perspectives. Or, to phrase the point differently: one of the main functions of ED 2000 is to renew the debate on the relationship between the traditional institutions of participation and co-operation on the one hand and the globalization processes on the other. This demands a new conceptual apparatus as well as 'new facts' to be developed in interaction with each other through a dialogue between all concerned.

It is possible to muster an argument against 'agreement on words' as a precondition for objectivity on the part of research: by refusing to agree on definitions any person or institution among those concerned can prevent anyone from assuming an objective role. Again, we face a complex problem. 'Agreement on words' is no simple process of voting, where anyone can block progress by voting against a certain interpretation of, say, democracy. In an open dialogue where all concerned are obliged to listen to each other, consider arguments and pursue the most reasonable interpretation, it is not up to all actors to individually decide what to opt for unless they want to opt out of the development coalition at the same time. But then they no longer belong to those who need to be considered in assessing when a linguistic platform for an autonomous role for, say, research is present. Only by accepting the rules of a 'democratic dialogue' — or whatever we choose to call it — is it possible to exert influence over other actors in a development coalition.

The need to consider 'linguistic tools' is well illustrated by the following contributions, where each of the research groups sets out its line of thought and action. This is done in words — in a sense the road to the workplace is paved with words. The words are of many different kinds and refer to different 'phenomena' — theory, cases, experiences, knowledge. When brought together in a particular pattern they identify 'a position' — a particular selection of problems, enterprises, points of departure, patterns of co-operation, results to be achieved. While the various groups 'touch base' here and there, they also differ. Mainly, however, the differences do not express themselves in terms of different and conflicting perceptions of 'the same reality'. Rather, the main source of differences is that the realities to which the groups relate are described differently — the linguistic tools applied to identify enterprises, problems and approaches are not identical. The development of general views — or general theory — is more an issue of bringing the different linguistic universes to convergence than it is to reach simple agreements on 'what constitutes the truth'.

As a point of departure, the relationship between the linguistically established interpretations and approaches of different researchers or research groups can be of different kinds.

The first possibility is that they compete: they identify the same reality as the focus of interest but give it different interpretations and suggest different forms of action. This situation has tended to play a dominant role in theory of science debates and a concept such as 'paradigm' is generally used to understand why 'the same reality' can be seen differently.

Much more common, however, is disconnectedness: the definitions of reality are different, interpretations differ and suggestions for improvements differ, too. Even if the reality is in some way or another the same — i.e., 'the enterprise' — the way in which it is conceptualized and interesting aspects identified differ, and they differ in such a way that they do not touch base.

Weak connectedness is a further alternative: One group wants to deal, for instance, with questions of leadership while another wants to deal with man-technology interrelationships. However, those interested in leadership may advance views on leadership that can support certain structures on the production floor, or have a disruptive effect. Likewise, the shop floor group can argue approaches that give preference to one kind of leadership rather than another. In this way they can touch upon each other's territory, but only in a weak sense. A discussion between them will, as a point of departure, not be very fruitful since those areas where they touch are small compared to those where they do not touch, at the same time as the position of each group will hang much more on the totality of its views and approaches than on a partial issue such as the meeting ground exemplified above.

The purpose of ED 2000 is to achieve strong connectedness; although the right of each group to pursue its own problems and issues through its own approaches is not to be questioned, the total landscape emerging under ED 2000 should resemble a 'quilt' — a landscape of numerous and strong interfaces. In this way the total landscape can be set in motion and not only each element. When strong connectedness and 'quilting' is to be preferred to competition it is because the historical experiences with competition and associated wars over theory and 'paradigms' are not positive from a developmental perspective. When each research group, in line with the tradition of critical theory, has chopped the head off at least one other group, all groups will be one head shorter — to the great glee of all evil-minded spectators but hardly anyone else. The effort to combine autonomy with numerous interfaces and ongoing discussions is the background to the choice of the word 'module' to describe the status of each unit within the program. A 'module' is a unit on its own, with its own characteristics and functions but also with a role to play within a larger whole, where it is this larger whole that gives much of the meaning and potential to each of the units.

The project chaos

Often when reading even contemporary case studies of organizational development one gets the impression that, although general theory may have lost much of its persuasive force, when a local theory of organization is developed, it is full speed ahead. This presupposes that there exists one identifiable stream of events that constitutes 'the organizational transformation of the enterprise' and one specifiable understanding that 'controls' this stream of events. As will emerge from the presentations that follow, however, this 'single track transformation process' is seldom seen. If anything, each enterprise is today characterized by trying to run a *number* of improvement processes that tend to overlap: some may be technological, some may pertain to IT, still others may pertain to benchmarking, and so on. They can all, however, easily have implications for, say, work organization. At the same time they often imply different views on the whys and hows of change. The contemporary 'post-modern' drift towards a pluralist approach to theory of organization and strategy for change is not limited to research circles. Rather, exactly the same pluralism is found all the way out to the workplaces.

In most cases the somewhat chaotic project portfolio encountered reflects a transitional phase; a phase between a past when change and 'projects' were a rarity and a future where what is today dealt with in 'projects' — change and development — will be the main activity of the organization. Today, however, a program like ED 2000 seems to 'compete' with large portfolios of other projects. There is often a certain 'project weariness' present: in the enterprises there is so much 'going on' already that it is difficult to create enthusiasm for still 'another project'. However weary people are of projects and however difficult it is to gain a comprehensive view of what is going on, they can get no consolation other than being told that it will get worse before it can get better. Better it will be, but only when improvement and innovation have become the core activity for all actors and when they are no longer seen as a secondary activity, taking time away from 'the real job'. It is a major aspect of ED 2000 not only to relay this message but also to help the enterprises on their way towards a strong organization of continuous improvement. In this way ED 2000 makes a conscious effort to transcend the situation that often prevails today to help the enterprises move beyond 'project thinking' to embark on processes of broad and permanent development where all main issues and areas are covered. In this way ED 2000 emerges as a set of 'meta-projects', helping enterprises to organize and run projects, rather than a new family of projects.

This point leads to another, namely that organization for development — or development organization as it is also called — is becoming a continuously more important part of the concept of 'organization development'. The traditional focus on what to do in ongoing production is not fully replaced by a focus on how to master change as a permanent activity, but this last issue is continuously growing in importance.

The content of the book

The first of the following contributions has its origin in the most recent module to be established under the program, located in Agder in the south of Norway. The contribution reflects the considerations and steps needed to establish a module. Departing from a case, it also introduces 'the globalization issue' in a particular form — the problem that it is always cheaper to produce somewhere else. As long as cost efficiency is the only criterion, many workplaces in Scandinavia will live under a constant threat of closure. When a particular plant was threatened by closure the debate did, however, move into another channel, focusing on competence and more innovative contributions to the mother enterprise. With this shift in frame of reference it became possible to save the plant since innovative capacity is dependent on a far more complex set of factors — many of them not easily transferable — than the factors involved in allocating production purely on a cost efficiency basis. With this, a way to save and expand on workplaces through the development of capacity for learning and innovation is also pointed out.

In the next contribution, emanating from a module shared between the Norwegian School of Economics and Business Administration and the Trades Unions' Research Institute in Oslo, the relationship between the global and the local is pursued further, in particular in terms of the relationships between external pressures confronting the enterprises on the one hand and the traditional Norwegian/Scandinavian patterns of labor-management co-operation on the other. The contribution focuses on the co-operation councils, a core formal body in the participatory system. It is demonstrated that such bodies can be linked to at least three main processes in the enterprise: controlling ongoing production, arenas for labor-management negotiations and, thirdly, playing a role in development. It is shown that the demands placed on the councils in terms of role configurations and forms of co-operation differ between these functions, and that the external pressures today demand an expansion of the development function and its corresponding set of role patterns. Ways are outlined by which this can be achieved in practice.

In the third contribution — from The Work Research Institute in Oslo — the nature and demands of development processes are further pursued, with an emphasis on the idea of development as a task *sui generis*, demanding development organization as analytically different from work organization. While the main tool in ongoing production is physical resources, the main tool in development processes is communication. Communicative instruments are presented and discussed together with a need to make participation in the development organization such that it can cover all members of the enterprise.

The fourth contribution relates to fishing, fish farming and fish processing, a major part of the Norwegian economy. Located within a module based on research groups in Tromsö and Bodö, the transformation of this sector from traditionalism and generally cheap products for selective markets into high quality products for the world market forms the background. In this kind of context, where a dramatic process of adaptation to new global conditions is taking place, many of the requisite processes need to be constructed almost from scratch and the multiplex role research needs to undertake is demonstrated, ranging from providing concrete support to specific processes on the level of the workplace to helping promote new understandings of basic concepts such as productivity and business strategy.

The fifth contribution — from a module located at the Norwegian University of Science and Technology in Trondheim together with an associated foundation for applied research (SINTEF) — deals with the interplay between technology and organization. This is a classical topic, in the sense that numerous efforts have been made to see how technology and organization can be made to structurally match each other. This was, for instance, a core element in the socio-technical school as it originally emerged in the 1950s. In this contribution the point of departure is different: the match has to be processual: it has to be worked out in each specific local context as part of ongoing development processes. There are no principles that can be universally applied. The task is rather a ‘constructivist’ one. In this way the paper argues that the solution to the classical problem of how to create the optimum man—machine interaction is developmental rather than a matter of mere definition. It turns the focus on the kind of research that can be involved in specific workplace processes and outlines how the Trondheim module works with a number of cases.

The last two chapters focus on networks. Beginning with Nordvest-Forum, what is probably the most substantial network for learning and development in leadership and organization in Scandinavia is introduced. Consisting of about 50 enterprises with a total of about 10,000 workplaces, the network was formed in the latter 1980s and has been in existence now for a number of years. The contribution highlights how enterprises can work together and use each other to broaden the scope and depth of the experiences available to each of the members and, through this, increase the potential for experiential learning. It also demonstrates how experiential learning should go together with exposure to more general concepts and theories to ensure a continuous inflow of ideas that can break up the often stalemated local understandings. 'Globalization' is not only a question of competition, it is also a question of the development and circulation of flows of ideas within fields like quality and logistics, and of the need to bring these ideas into confrontation with 'the local understandings' that often get a solid grip on actors in the industrial periphery. It is, however, also emphasized that general perspectives have to work through merging with local understandings to form new wholes — a general view in itself can never be 'applied' locally.

In the last contribution, from Rogalandsforskning (The Rogaland Research Foundation), we actually confront three networks, all of them quite major and spanning small suppliers up to some of the main national actors in the offshore oil industry. Again, the focus is on new formations emerging because of the new conditions for competition: on the one hand the relevant environment becomes global, on the other hand 'the local' is also gaining in importance through processes where enterprises join forces to deal with the challenges together. One of the networks is probably the oldest one in existence in Norway, dating back to the 1950s. This network demonstrates some of the dynamics of network cooperation itself, often starting with such issues as joint production and eventually ending up with broad development tasks spanning all major levels and issues within the enterprise.

When enterprises join forces to create learning contexts for each other, we are beyond stable structures and into a dynamic notion of 'network'. Each specific configuration of members, relationships and activities is transitional. The complexity can easily get high and the futility of 'steering' network processes through some uniform understanding or other that can give direction to all the participants and activities becomes very clear. Learning networks represent moving fields where only a continuous work with positionings and relationships can achieve the goals of the network. The need for research to be a co-actor rather than a spectator is very explicitly demonstrated.

Living with pluralism

The idea of research has traditionally been associated with identifying how phenomena are, preferably in terms of invariant characteristics. 'Theory of organization' is no exception, and quite a number of such theories — in terms of individual efforts or broader schools of thought — have seen the light of day: the human relations school, the socio-technical school, the labor process school, the institutional school, and others. Numerous concepts have been launched to indicate a general path to the best pattern of organization: organic organization, joint socio-technical optimization, lean production, business process re-engineering, total quality management, learning organization, and others. Enterprise Development 2000 departs from the premiss that all these concepts, ideas, theories, schools of thought, or approaches to improvement, have something to offer — in fact, so have a number of other ideas that remain unmentioned here. As soon as we abandon the notion that a single conceptual scheme can claim to successfully control the field, a whole new generation of issues start to emerge. Living with pluralism does not mean that one has to become the complete relativist and leave it at that: living with pluralism to the benefit if all is instead quite difficult. It means that we have to focus on how we learn from each other, how we position experiences in alternative figure—ground relationships, how we establish flows of communication and arenas for discussion. Rather than think in terms of the best theory, we need to think in terms of optimum flows of theories; rather than think in terms of who is best we need to think in terms of how we can all be better. The main purpose of ED 2000 is to attack these problems, in practical terms, but with research as a partner; to see what challenges they pose and how to deal with them. In this sense ED 2000 is a learning expedition, but what is to be learned is not new theory of organization, nor even a theory of organization development, but how to create patterns that make us all able to learn from differences.

References

Argyris, C. & Schøn, D.A. 1978. *Organizational learning: A theory of action perspective.* Addison-Wesley, Reading, Ma.

Fals Borda, O. 1987. The application of participatory action research in Latin America. *International Sociology*, Vol. 2, 329-347.

Giddens, A. 1990. *The consequences of modernity.* Polity Press, Oxford.

Greenwood, D.J. & Santos, J.L.G. 1992. *Industrial democracy as a process: Participatory action research in the Fagor co-operative group of Mondragon.* Van Gorcum, Assen.

Guba, E.G. & Lincoln, Y.S. 1989. *Fourth generation evaluation.* Sage, Newbury Park.

Gustavsen, B. 1992. *Dialogue and development.* Van Gorcum, Assen.

Gustavsen, B. 1993. Creating productive structures: The role of research and development. In: Naschold, F. et al.: *Constructing the new industrial society.* Van Gorcum, Assen.

Gustavsen, B. 1996. Development and the social sciences — an uneasy relationship. In: Toulmin, S. & Gustavsen, B. (eds): *Beyond theory: changing organizations through participation.* John Benjamins, Amsterdam/Philadelphia.

Gustavsen, B., Hofmaier, B., Ekman Philips, M. & Wikman, A. 1996. *Concept-driven development and the organization of the process of change.* John Benjamins, Amsterdam/ Philadelphia.

Gustavsen, B. 1997. Is theory useful? *Concepts and Transformations,* Vol. 1. No. 1, 63-78.

Isaacs, W. 1993. The dialogue program at the Center for Organizational Learning. Work note.

Lewin, K. 1947. *Die Løsung Sozialer Konflikte*. I.M. Christian Verlag, Bad Neuheim.

Naschold, F. 1993. Organization development: National programs in the context of international competition. In: Naschold, F. et al.: *Constructing the new industrial society*. Van Gorcum, Assen.

Reason, P. 1993. Sitting between appreciation and disappointment: A critique of the special issue of Human Relations on action research. *Human Relations*, Vol. 46, No. 10, 1253-1270.

Schwandt, T. 1996. End of criteriology? *Qualitative Inquiry,* Vol. 2, No. 1, 36-52.

Toulmin, S. 1990. *Cosmopolis. The hidden agenda of modernity*. The University of Chicago Press.

Whyte, W.F. (ed.) 1991. *Participatory action research*. Sage, Newbury Park.

Chapter 2

Mentor Leadership and Theory Building in Practice

Harald Knudsen

The Agder region is a newcomer to the ED 2000 project. The 'Southern' or 'Agder module' started in October 1996 and seeks to initiate and develop individual and organizational learning processes in participant enterprises and in associated networks of firms. The project is sponsored by two large networks of local firms, the 'Kompetansering Sør' (28 firms, mainly located in West Agder) and the 'IT-ringen' (45 firms, mainly in East Agder). However, the project also includes companies that are not members of the networks. At the time of writing, specific schemes for development cooperation are being created with 7–8 firms.

The first part of this chapter presents an outline of the main strategic challenges facing the project team in our region. The second part outlines a program of training and action learning.

1. A case in point

In December 1996 the Swedish Ericsson concern, an international IT giant of 95,000 employees, operating in more than 120 countries, announced plans to move its Arendal plant in the Agder region to a new high-tech park at Fornebu, Oslo. More than 400 jobs were at stake in Arendal, as was the role of the Ericsson plant as a locomotive of the regional IT business. The main argument for moving the plant was that the concentration in Oslo of all Ericsson activities in Norway would yield greater efficiency. It would bring the company closer to the main customer, the Norwegian telephone company, Telenor. It would also stimulate technological development and innovation, since the Oslo area has the greatest concentration of information technology research and development in Norway. Thus, moving to Oslo was considered to be necessary in order to secure the future operations of Ericsson in Norway.

Ericsson's radical plans caused a great deal of local protest: from Ericsson employees and their families, from local politicians, from nearby academic institutions, and from the business community in general. The main argument against moving the plant was that the employees did not want to move. Corporate managers had calculated that 80 percent of the employees would have to move to Oslo, in order for the strategy to be successful. First of all, it was argued that very few families wanted to move, and that they most likely wouldn't move. Also, it was argued that the wage level and the cost of living in Arendal were lower than in Oslo — which would save expenses for the company. Furthermore, the Arendal plant had a lower turnover of employees, and lower costs of training, than a sister plant in Asker, outside Oslo. The recruitment of new personnel would be easier in Arendal than in Oslo — or, it would at least be an advantage to recruit in both locations. Against the arguments in favor of concentration and closeness to customers and to major IT-centers, it was furthermore argued that a main benefit of modern IT is that you can communicate without physical-geographical concentration. It was pointed out that there was a 'logic of efficiency' in retaining the plant at Arendal.

In the end, the local forces prevailed. The decision to move was reversed. Local politicians promised to improve the regional communications infrastructure for automobile and air traffic, and to lower the rents for factory space. Local management promised to work hard to improve efficiency and to establish effective links of cooperation with the Oslo facilities.

The Ericsson case is indicative of the kind of forces that are at work, surrounding the traditional meeting places of owners and workers, suppliers and customers. Thirty years ago, when talking about enterprise development, there was seldom any reference to corporate strategies. Regional considerations might have played a role in developmental thinking, but mainly in terms of the exploitation of natural resources, the availability of physical infrastructure, labor supply, and governmental subsidy schemes and protection. The problem to be overcome was the conflict between capital and labor.

Much of this has changed today. While there is still a tension between capital and labor, it is not so much over the distribution of wages and profits as over business strategies and strategic positioning. While regional infrastructure is still important, there is less focus on the general factors of production, and more on those factors that are decisive for learning and logistics. And while Enterprise Development (ED) is still about participation and communication, the content of both has changed. While internal groups and teams still occupy a central place, there is a greater influence of international management concepts, of corporate policies and strategies, and of internal and external, local and global, network structures.

Globalizing forces

Globalization involves two major trends. On the one hand, the textbook case of globalization points to the global standardization of products, the global unification of quality standards, and the global minimization of costs of production and marketing — including manufacturing, transportation, product development, services, and logistics. In the extreme case, each subsidiary, and each unit in the value chain of a global network, can be considered a 'castrato'. It has been 'downsized' and stripped, of any function that is not cost-optimal or part of the 'core' business. As such, it is not properly a profit center, but rather a 'cost center'. Unlike previous generations of 'multinational' or 'international enterprises', where the local goal was to maximize profits, a predominant strategy of 'global enterprise' is integration, standardization and cost rationalization at each separate point in the chain (for a review of trends and strategic options, see Bartlett and Ghoshal, 1992).

One of the consequences of global cost rationalization is that jobs become insecure. Sometimes, machines and complete lines of production are moved to different parts of the globe at short notice. Units of production, which are profitable by local standards, may not suit the global strategy of the firm, and will be laid off. Local subsidiaries may, by global strategic design, be cut off from product, process or market development. Such companies tend to have a minimal developmental effect on the host environment, and they are maximally vulnerable to the whims of global strategies.

The rationalization trend represents only one side of the coin, however. There is another dimension of globalization which goes in the opposite direction, which is concerned with maximizing global learning. Companies maximizing their global learning potential are sometimes called transnational enterprises. The creation of new knowledge, and the spreading of ideas, of 'best practices', and innovations on a global scale, are as much a part of globalization as is cost rationalization.

The global learning ambition tends to make rationalization strategies more negotiable and to redefine the relationship between each unit in the chain and its host environment. Units that are capable of supplying technology and innovation to the rest of the concern tend to gain strategic power. Innovations are not easy to predict, and global firms cannot simply transfer innovative capacity. Neither can they easily transfer knowledge, since much knowledge tends to be tacit. Furthermore, sources of information, and stimuli to innovation, are frequently found in both internal and external environments. Both types of environment can be tapped for information, but neither type can be geographically moved without great costs. The more integrated the subsidiary is in local supply systems and in local innovative environments, and the higher the quality of the local workforce, the more robust is the political stance of the subsidiary.

Regional forces

In the Ericsson case, the employees and their families sided with local management and local politicians against corporate management. The overarching local interest was to secure long term, stable growth. Ericsson asked for local improvement of the communications infrastructure and for lower costs of production facilities. From the point of view of global cost rationalization strategies, such demands are to be expected. However, from the point of view of securing long term, local bargaining strength, meeting such demands is not sufficient to secure the local interest. There is no fun in a high-cost culture trying to compete with low-cost countries around the world on a *cost only* basis. The way a high-cost location can compete with low-cost locations is rather by adding value through learning and innovation. And the only way the local work force, local management and local politicians can move strategically to enhance the long term growth prospects is by stimulating not only the innovative capability of the plant, but also the surrounding learning environment.

A salient feature of such local learning environments is the coming together of a business interest and a learning interest in an atmosphere of openness and trust. Preferably, not all participants should be global subsidiaries. Small, locally owned firms may play an important role as facilitators and as models of entrepreneurship. While advanced, specialized factors of production are important (Porter, 1990), the most important learning element is the implicit, experiential learning that takes place in the factories, in the markets, and at informal gatherings (Marsick and Watkins, 1990). In progressive learning environments there is a collective learning process going on, where the free flow of information, even among competitors, is an ingrained part of formal and informal learning networks.

Thus, there seems to be a complex combination of possible strategies surrounding local ED processes. Global firms may pursue global rationalization strategies or global learning strategies, or combinations of these. Local and regional interests may pursue defensive 'factor' strategies or offensive 'innovation' strategies, or combinations.

The worst possible combination, from the point of view of the local environment, is to pursue a *defensive factor cost strategy* to attract global firms pursuing *cost rationalization strategies*, but ending up with a plant closure — since there will always be some place in the world where labor and natural resources are cheaper. The best possible case is where *offensive learning-environmental strategies* attract the attention of firms pursuing *global learning strategies*. The company will then want to locate in the region, which in turn will have a positive influence on the existing networks. A positive cycle may then develop, which will in turn attract other investors.

Caught in the middle are employees and managers of local subsidiaries and of local firms that are dependent on such subsidiaries, both for business and for learning purposes. The local subsidiary can play an active role, mediating between the *global push* and the *local pull*, both on a unilateral basis and in cooperation with local firms.

Local ED is to a great extent a matter of combining the strengths of management and workers in a united fight for survival against global competitive forces. It is a matter of meeting the goals set by national and global concerns, while at the same time exploiting the benefits of national and global sources of technology, thereby reinforcing the regional learning-environmental pull.

This perspective is relevant to a great number of firms in the Agder region. Either the major companies are subsidiaries of global concerns, or of Norwegian concerns pursuing global strategies, or they have business or information links to such subsidiaries (as suppliers, customers, consultants, neighbors, informants, etc.). It is therefore mandatory that the ED 2000 project takes these forces into consideration.

2. ED 2000 in the Agder region

Since the Agder module is the most recent addition to the ED 2000 project, and is just in the process of getting started, it would be premature to present details on research methods and intervention procedures. However, the basic scheme is as follows.

The Agder module aims at three different levels of contact — or 'intervention' — with participating firms. First, at any one time, we expect to have 6–8 firms participating in *intensive development programs*. At the time of writing (June 1997), program schemes are being worked out with participant firms. Second, through the two networks mentioned above, and through contacts with other projects of the Agder Research Institute, we expect to have a varying number of firms participating in *network development programs*. Finally, throughout the period of the program, there is an ongoing (academic) research effort, mandating off-and-on unobtrusive observation, here called a *formal research program*.

The intervention profile of the Agder module

The idea of workplace and enterprise development through cooperation between researchers and practitioners is not something that is easily understood by companies. Although we had indirect contact with a great number of firms in the two networks (Kompetansering Sør and IT-ringen), and even though 25 firms participated in our 'Start conference' at the beginning of February, only one company has come forward 'voluntarily' and asked to be one of the 'intensive development program' firms. In all other cases, agreements have only been reached after one, or even several requests. Many companies of the 'local subsidiary' type, are interested in participation, but so caught up in corporate renewal plans, reorganization schemes and company-wide training programs that they simply cannot take more 'talk'. In other companies it may seem that managers are hesitant, because they are afraid of losing power — they are balancing on a fine edge between corporate strategies and local demands for survival. Still others may be generally skeptical of researchers and research institutions.

Whatever the reason, and whatever the situation in other regions, clearly, in the Agder region, it has been necessary to make a selling effort. Also, it has been necessary to be reasonably clear about the 'product'. The companies, in particular the management side, want to be reasonably well informed about what to expect. At the same time, however, they are also very eager to make modifications and changes to the 'product', as soon as they get their hands on it.

In the case of intensive development programs, companies are expected to pay half the cost of a 'standard package'. In addition, the companies will bear the investments in work time and other expenses. Such a standard package includes four main elements, described below.

1. *Initial Process*

The standard package starts with preliminary interviews and a one-day 'dialogue conference'.

The initial interviews are considered quite important. Although the dialogue conference may be a good opportunity to discuss major problem areas and alternatives, it is not always a good occasion to bring out emotions, secrets and gut reactions. The dialogue conference is a 'public' setting. Some participants hesitate to talk about sensitive issues in such a setting, but they may want to do so in the privacy of an interview. Also, for the BU team, it is an advantage to have preliminary knowledge of problems, ideas and hopes, before staging the dialogue conference.

For the content of the preliminary interview, we have devised three possible alternatives. Some companies have very definite plans and ambitions for development. The initial interview will then be focused, but open-ended. Other companies are 'snoozing' — they have no definite plans, and they are uncertain about attitudes and feelings. In that case, the initial interview will be broad and open-ended. Finally, some companies are aware of ideas and impulses, but they are groping for facts about the present situation and about perceptual differences between groups of participants. These are offered a more structured interview scheme, including a quantitative analysis of results. In all three approaches, the role of the researcher should be one of active listening (the latter type of interview calling more for recording than for active listening, however).

The objective of the dialogue conference is to gain acceptance for a continued process, decide on the further direction and sequencing of the process, and on matters of team composition and team work. The role of the researcher should be mainly one of facilitating the process, not disregarding direct questioning and recommendations when necessary, however.

From there on, the project team will work in three different areas of intervention:

2. *Installing Barometers*

The Agder Research Institute has considerable experience with installing 'barometers' (or 'gauges') to measure employee satisfaction/employee loyalty and customer satisfaction/customer loyalty. As part of the total package, all participant firms are offered the free installation of two such 'barometers'. The idea is to increase the awareness of the link between internal culture and customer loyalty, both among managers and employees.

The role of the researcher here is one of being an expert consultant.

3. *Leadership Training*

Leadership training is not only for managers, not even mainly for managers. Rather, it is for all levels of participants — for all the main actors involved in the ED process. Training in team work and interpersonal communication will receive special attention. This also includes learning about 'learning-to-learn', both as related to technical skills, to personal traits and abilities, and to creative skills. As part of the standard package, participant firms are offered two (two day) training seminars in leadership — broadly defined — and one more specialized (one week) session for a specialized topic (the choice includes internationalization, women in leadership, innovation, mentoring, or special training as part of the implementation of more 'standard', international concepts of management, such as TQM or BPR). In the training sessions the team of researchers should play many different roles, such as lecturers, process consultants and dialogue partners.

4. *Change Agent Intervention*

The nature of intervention and the definition of the change agent's role obviously depends on what comes out of the initial interviews and the dialogue conferences. The philosophy behind the project would favor 'action learning' approaches over traditional 'organization development', but most likely the intervention will include both of these and some others. An overriding theme will be 'theory building in practice' (Knudsen, 1997b), relating both to individual learning and growth, technical innovation and the cultural development of the firm.

These elements, then, all belong to the standard 'package' offered to firms participating in the intensive development programs. Let me, however, restate that the mission of such a 'product package' is not to shortcut the dialogue of ED, but only to establish a practical basis for planning and for cost sharing between the research institution and the participating firms. The package offered will probably in no case be used as it is. All our intervention programs aim at further dialogue and reciprocal learning.

Let us now consider the two other parts of the program, the network development programs and the formal research programs.

Network level interventions

In a number of problem areas, leaders would like to change a culture directly and instantly. Politicians, unions, interest groups or business owners would like to have formulas for instant change. Owners might want to develop instant creativity or instant 're-engineering'. Women's organizations might want to install instant equal rights. Employee organizations might want to have instant open communications. And proponents of change would like to change the culture directly, without having to work with individual men and women, through mentoring, counselling, leadership training or team efforts.

On the one hand, no social scientist can or should produce such formulas for instant change. On the other hand, we do by now know more about experiential learning, about change and about organizational cultures than we did only a few years ago.

A main ingredient in experience-based innovative development is *divergent thinking*, individually and in teams (for a brief review see Woodman, Sawyer and Griffin, 1993. More elaborate presentations are found in Knudsen 1998a and 1998b). Divergent thinking is a direct result of making use of one's *variety of experience*. The individual utilizes variety of experience and divergent thinking to make *innovations* and to seek new *applications*. An *innovation* means to combine what is known (variety of experience) in new ways. *Applications* means to take what is already known and apply it in new contexts.

At the group or team level, divergent thinking is a product of individual variety of experience and *intimacy of communication*. High-variety groups have greater potential for creative thinking, but they usually have greater problems of communication. To the extent that high-variety groups are capable of effective, open communication, they have a strong potential for development.

An intervention strategy geared towards experience-based performance improvement therefore needs to consider both the variety element and the communication element. Individual variety of experience can be stimulated by opportunities for rotation, travel, training, reading, experimental work, and all kinds of interpersonal contacts, within and outside the firm, in the same profession and in other fields. Group variety can be increased by putting broadly experienced people together, but also by putting *different* people together, from inside or outside the organization. In order for this to produce creative thought, however, effective communication is mandatory. And effective communication depends both on communication skills and on the general atmosphere of trust.

Learning in networks will be facilitated in the Agder module, by use of several, different strategies. An intranet is being set up, where the main idea is to write down, collect, and distribute the 'good stories' and the whereabouts of how to 'embody' them. While the contract firms are seen as leading agents of change in the region, the diffusion of results to other parts of the business community will be done mainly through the two sponsoring networks, but also through informal contacts, overlapping team membership, and — not least — through direct information about results as the project goes on.

The role of the researcher in all of this is mainly that of being a practical caretaker, informant, and facilitator.

Formal research

The shaping of the national ED 2000 project owes much to a long-standing Norwegian tradition of cooperation between the major actors of the labor markets, and between social science research and working life. Such a tradition is, implicitly or explicitly, part of all the modules of the ED 2000 project. However, it should also be recognized that there is no one theory of enterprise diagnosis or intervention behind the project, neither at the national nor the regional level.

A major research effort in the Agder module will be directed toward the *constitutional and innovative* aspects of ED, combining structural, processual and learning elements of development.

An important *constitutional* issue, linked directly to the problems of globalization mentioned in the introduction, is the extent to which the traditional meeting in the factory, between labor and management/owners, is a 'fake' in the global firm. Is the traditional 'meeting of the parties' an authentic arena for dialogue and decision? Or is management so strongly caught in the webs of globalization that the meeting becomes little more than a ritual — where important matters are always decided somewhere else? More specifically, what role can such cooperation play under various forms of globalization?

At a more general level, the constitutional issue raises some fundamental problems of moral philosophy and pragmatism. In the meeting between individuals and enterprise, and between enterprise and environment, what is the relationship between tacit and explicit, individual and organizational learning, and what is the relationship between learning and enterprise constitution? Is there anything like an 'ideal regime' of enterprise development (Knudsen, 1998a)?

Enterprise development of the ED 2000 type means something more than just product innovation or productivity increase. It is the creation of open lines of communication, of relations and team work marked by trust and respect, that is at the forefront. This is the essence of what I call *mentor leadership*. The relationship between such *mentor leadership*, on the one hand, and the resulting innovative and productive capacity, on the other, is to be critically examined. Interfering with this relationship are a great number of sometimes quite paradoxical structural variables. An innovative climate may need both *top down* and *bottom up* impulses, calling perhaps for a *middle-up-down* strategy of development (Nonaka, 1994, Nonaka and Takeuchi, 1995). The process may need both freedom and control, openness and structure (Kanter, 1988). According to Pascale (1990), too much energy has been devoted to maintaining equilibrium, eliminating tensions and enhancing consistency, thereby sacrificing *vitality*. Not even the commitment to a consistent ideology is seen as being unequivocally positive for the company. Too strong a commitment in any one direction may produce vulnerability to sudden environmental shifts. Or, citing Miller (1990, p. 18): "overconfident, complacent executives extend the very factors that contribute to success, to the point where they cause decline".

Such paradoxes — and many more could be cited — may offer an explanation of why studies of 'the benefits of participation' have offered only inconclusive (although mainly positive) evidence concerning the relationship between cooperation and performance. A more profound understanding of these issues requires attention both to the paradoxical structural variables and to the phenomenologically fine-tuned elements of leadership, cooperation and team work. The number of meetings and joint committees may not be a good indicator of *mentor leadership*.

These two broad areas, expressed by the key words *constitution* and *innovation*, will be the subject of basic research in the Agder module — producing, we hope, two or three doctoral dissertations in addition to scholarly articles. Related to these areas, there will also be associated research projects concerning the roles of women in leadership and in organization cultures, and concerning issues of internationalization and organizational learning.

Summary

A great number of enterprises in the Agder region are 'caught in the middle' between corporate interests and local interests — between a 'global push' alternating between cost rationalization and learning strategies, on the one hand, and a 'regional pull' alternating between defensive factor-cost strategies and offensive learning-environment strategies, on the other. Furthermore, even if local companies are not 'caught in the middle' to the same degree, they nevertheless frequently depend on global subsidiaries for contracts and innovative stimuli.

This overarching political-economic framework influences the Agder module of ED 2000 in several ways. While 'lean production', cost rationalization or 'downsizing' will certainly remain with us, a major goal is to enact local learning processes countering the global threats. In these learning processes there is a close tie between innovation, experiential learning and the concept of democracy, including work democracy. The basic requirement of democracy is that people who are affected by decisions should have a right to participate in those decisions (Dahl, 1985). The main secret of experiential learning is that one and the same person, or one and the same team, are entitled both to think and to act. To be a *fully fledged member* of the organization means to be voluntarily, fully involved in both thinking and acting. This is a prerequisite for work democracy, and it is also the key to experiential learning, theory building in practice, and mentor leadership.

The principles of democracy, learning and mentoring attitudes form the basis for the project. The three major components are the *intensive development programs* (including 6–8 firms at any one time), the *network development programs* (including firms that are already members of two formally organized networks), and the more academic *formal research program*. The essence of the intensive development program is to work directly with the companies to effect changes and improvements, appealing to the global learning strategies of the firm. This is done, *inter alia*, by developing mentor leadership and by engaging the workforce in theory building in practice. The essence of the network development program is to add to the dynamism of reciprocal learning, stimulating the offensive, regional learning-environment strategy. And the essence of the formal research program is to add to our insights about contexts and constitutions, about cooperation and authenticity, and about innovation, learning and leadership.

References

Bartlett, C.A. and Ghoshal, S. 1992. *Transnational Management: Text, Cases, and Readings in Cross-Border Management*. Homewood, Ill.: Irwin.

Collins, J.C. and Porras, J.I. 1995. "Building a visionary company". *California Management Review* 37(2, Winter): 80–100.

Dahl, R.A. 1985. *A Preface to Economic Democracy*. Berkeley/Los Angeles: University of California Press.

Ghoshal, S. and Bartlett, C.A. 1994. "Changing the role of top management: beyond strategy to purpose". *Harvard Business Review* November–December: 79-88.

Ghoshal, S. and Bartlett, C.A. 1995a. "Changing the role of top management: beyond structure to process". *Harvard Business Review* January–February: 86–96.

Ghoshal, S. and Bartlett, C.A.. 1995b. "Changing the role of top management: beyond systems to people". *Harvard Business Review* May–June: 132–142.

Heskett, J.L., Jones, T.O., Loveman, G.W., Sasser, Jr., W.E., and Schlesinger, L.A. 1994. "Putting the service-profit chain to work". *Harvard Business Review* March–April: 164–174.

Knudsen, H. 1998a. *Reisen for* theoria's *skyld* (In Search of Theoria). Kristiansand: Norwegian Academic Press.

Knudsen, H. 1998b. *Teoribygging i praksis* (Theory-building in Practice). Kristiansand: Norwegian Academic Press.

Marsick, V. and Watkins, K.E. 1990. *Informal and Incidental Learning in the Workplace*. London, New York: Routledge.

Miller, D. 1990. *The Icarus Paradox: How Exceptional Companies Bring About Their Own Downfall — New Lessons in the Dynamic of Corporate Success, Decline, and Renewal.* New York: Harper Business.

Nonaka, I. and Takeuchi, H. 1995. *The Knowledge Creating Company*. New York: Oxford University Press.

Nonaka, I. 1994. "A dynamic theory of organizational knowledge creation". *Organization Science* 5(1): 14–37.

Pascale, R.T. 1990. *Managing on the Edge: How the Smartest Companies use Conflict to Stay Ahead.* New York: Simon & Schuster.

Porter, M. E. 1990. *The Competitive Advantage of Nations*. London: Macmillan.

Weick, K. 1995. *Sensemaking in Organizations*. London: Sage Publications.

Chapter 3

Corporate Efficiency and Employee Participation

Tom Colbjørnsen and Eivind Falkum

Introduction and overview

This paper explicates the approach of the 'FAFO/NHH module' within the 'Enterprise Development 2000' research program. One of the main topics in this program is to investigate how the Norwegian systems and traditions of employee participation and union–management codetermination may contribute to the international competitiveness of Norwegian firms. The unique approach of the 'FAFO-NHH module' is to analyze this issue by combining efficiency-related and participatory approaches to corporate development, and to synthesize economic, sociological and political science approaches to organizations.

As a business school, NHH has focused on strategic and organizational issues from a managerial perspective, and emphasized the economic aspects of organizational developments. The essence of an economic approach is that corporations should maximize their efficiency, which means that they should be organized so that they produce as much as possible, given the available resources (Milgrom and Roberts, 1992).

FAFO has usually done research within a participatory perspective, emphasizing the relationships between employee participation, corporate developments, union–management codetermination, and working conditions. The research has been performed within sociological and political science perspectives, rather than economics. The purpose of the FAFO/NHH module is to try to extract a synergy by combining these two approaches.

In this paper, we start by showing how corporate development in competitive markets is related to both efficiency and employee participation. Then we shall present two examples of our research so far — a study of communication in a globalizing firm that we conducted at the Norwegian pharmaceutical company Nycomed ASA, and an analysis of worker–management codetermination in

Siemens AS, a Norwegian subsidiary of the German Siemens AG. The former study illustrates an efficiency approach to corporate development, while the latter is an example of a participative approach. Finally, we develop a model that shows how participation and efficiency may be integrated into a single conceptual framework. The model has a dual purpose. First, it serves as an integrating framework for the different projects within the 'FAFO-NHH' module. Second, it integrates economic and sociological/political science approaches to corporate development.

Efficiency, participation and corporate development

Almost every sector of the economy is nowadays exposed to more intense global competition, implying a need for corporations to emphasize customer orientation, cost efficiency, and flexible organizational forms. Production is reorganized and relocated to reap economies of scale, scope and standardization, at the same time as market and product strategies are made responsive to customer expectations. Organizational forms must nurture and exploit unique competencies and other competitive resources, at the same time as they motivate and coordinate employee activities in a cost efficient way. Production processes and supply chains are improved. Mergers, acquisitions, alliances, networks, joint ventures, projects and teams are set up to achieve the joint goals of flexibility and market orientation on the one hand, and specialization, speed, uniqueness and managerial control on the other.

These organizational developments derive from the incentives that competitive markets give firms to use their resources as efficiently as possible (Colbjørnsen, 1997).

At the same time, however, Norwegian firms are part of a long tradition and history of employee participation, worker–management codetermination, and formal negotiations between unions and employer organizations. This history and tradition, often referred to as 'The Norwegian Model' (Dølvik *et al.*, 1997), has been strongly institutionalized in laws, formal agreements, procedures, and codetermination committees, as well as in informal norms and conventions.

There are several reasons why Norwegian firms might benefit from this system as they now try to adjust themselves to global competition. First, employees and their unions can invoke negative sanctions and conflict costs on firms that do not follow the institutionalized rules. Second, the system represents values of justice, democracy and participation that are held in high regard in Norwegian society. Firms that want to survive must to a certain extent incorporate the values of the society in which they are embedded. This point is stressed by the institutional approach to organization theory, which claims that any organization that is to survive should not only be efficient in an economic sense — it should also be consistent with important values in the society of which it is a part (see for example Scott, 1995). This holds for all kinds of organizations — including firms. Third, the system of influence and codetermination represents channels, procedures and processes for transmitting and developing information, competence, commitment and legitimacy between managers and employees. If the system works well, it may be an organizational asset that gives a competitive advantage for Norwegian firms, as it may stimulate employee motivation, foster employee–management dialogues that promote problem solving and creativity, lead to smoother implementation of organizational changes, and reduce conflict costs. Unique employer–employee relations that are embedded in history, traditions and corporate cultures, and that are supported by a nation's institutions and values, may be difficult for foreign competitors to imitate, and thereby confer a sustained competitive advantage (Barney, 1997). On the other hand, if the system does not work well, it may lead to obstruction, time-consuming political processes and conflict costs.

Communication in a globalizing firm: A pilot study of Nycomed ASA

Nycomed ASA is a pharmaceutical company with approximately 5500 employees. Nycomed's headquarters was until autumn 1997 located in Oslo, and the firm had a majority of Norwegian owners. In 1997 Nycomed was merged with the British corporation Amersham.

From having most of its operations in Norway and Scandinavia 10 years ago, Nycomed is now going through a process of intense globalization. This takes place through acquisitions and mergers. At the time of writing, the company has its own operations in 39 countries, while its products are sold in over 100 countries. Nycomed is involved in two business areas. Nycomed Imaging holds a global leadership position in diagnostic imaging products, and from this platform it seeks to expand into selective therapeutic areas relevant for hospitals. Nycomed Pharma provides affordable pharmaceuticals in selected product and market areas. This division has a strong presence in Europe, and is especially dominant in Norway and Denmark.

The philosophy of 'Enterprise Development 2000' is that the specific research topics should be developed through a dialogue between the researchers and the companies. The dialogue should single out areas that the companies regard as important in their development processes, at the same time as the researchers find them interesting and justifiable from a theoretical and methodological point of view. Another way to put it is that 'Enterprise Development 2000' tries to develop a new type of coalition or synergy between researchers and companies (Gustavsen, 1997).

The dialogue between Nycomed and the 'FAFO/NHH' module started in the summer of 1995, and the first topic that was singled out as mutually interesting was corporate communication in a globalizing firm. Nycomed has been through a rapid process of globalization, and was interested in discovering, in a systematic way, the kind of communication problems that may arise in the new global organization. Being a company that strongly emphasizes its corporate identity, Nycomed was especially interested in finding out what kind of communication could foster common corporate values and cultures in the new global organization. From the researchers' point of view, this topic was interesting because common identities and corporate values can be efficient integrating mechanisms in global organizations, where dividing national identities and subcultures may lead to time consuming and destructive conflicts. On a more general level, the resources spent on information processing, and the decisions made possible by the available information, are a strong determinant of corporate efficiency and rationality (see especially Stinchcombe, 1990).

It was decided to approach the issue of communication through an exploratory study in which the researchers combined insights from recent literature on strategic management and global organizations with interviews in the company. The respondents were selected from the headquarters in Oslo and two foreign subsidiaries in Europe, so that they would represent a variation in communication challenges. The interviews took place between November 1995 and February 1996, and a total of 32 respondents were interviewed.

It is important to stress here that the purpose of this first study was not to evaluate the appropriateness of Nycomed's global organization. Our respondents are not representative of the whole organization. Rather than estimating and testing the efficiency of the company's global communication system, the purpose was solely exploratory, in the sense that we wanted to systematize and identify those dimensions of corporate communications that seem most important to address from an efficiency point of view. The results that are reported below should therefore be read as a set of hypotheses about what aspects of communication are most important, and how they may influence the efficiency of the company's organization. In a later stage of our collaboration with Nycomed we plan to use part of the insights from this preliminary study to analyze how different types of work organization actually do influence communication efficiency, and how this may be changed through organizational development.

Based on the interviews and recent literature, we have classified four areas where communication appears to be especially important for corporate efficiency:

1. Market responsiveness;
2. Learning and innovation;
3. Production costs;
4. Corporate identity, coordination and management control.

As far as the recent literature is concerned, we have drawn especially on organizational economics (Milgrom and Roberts, 1992), resource based perspectives on the firm (Kay, 1993), core competence (Hamel and Prahalad, 1994), and global organizations (Bartlett and Ghoshal, 1991).

Before we elaborate on the results, it should be pointed out that the concept of communication is used in different ways in the literature. One aspect of communication has to do with how, and through what media, information is spread within and between organizations. Another aspect refers to dialogue processes, where actors interact in continuous problem solving. In this analysis, it is the first aspect of communication that is studied.

Market responsiveness

An important rationale for globalization is to gain access to larger markets, and have efficient distribution channels for the firm's product portfolio. However, globalizing firms may be tempted to give too much priority to product standardization in order to reach global economies of scale, at the expense of meeting specific product requirements in different regions (Bartlett and Ghoshal, 1991). To avoid this, the organization should have a communication system that enables it to be responsive to variations in local market conditions, and to exploit the increased product portfolio and market access that is made available through acquisitions and mergers.

- Information about variation in customers' needs, as well as competitive conditions and government regulations, must be transmitted from regional subsidiaries to production units. Such information is necessary in order to meet local variation in product requirements, either due to public regulations or customer preferences.
- Subsidiaries in different regions must communicate their product portfolios to each other in order to sell each other products. This may be especially important when a corporation is globalizing its operation through acquisitions, mergers or alliances with established firms who already have their own products. However, local units may have an incentive to withhold information about their product portfolio because they are afraid of losing control of product development, and because they fear that other units will get the honor and status that derives from being in charge of successful products.
- Information systems must support internal logistics, so that local subsidiaries can be supplied with the products they need at the appropriate time, either from central production units or other subsidiaries.

Learning and innovation

In recent literature on strategic management the core competence of the firm is claimed to be one of the most important sources of value creation and innovation

(Prahalad and Hamel, 1994). First, as core competence is unique to the firm, it may represent a competitive advantage. Second, since core competence consists of both formal education and the sum of the employees' experience, as well as firm-specific and intangible skills and insights, it is difficult for competitors to imitate. Therefore, core competence may give a competitive advantage that is more sustainable than a currently successful product portfolio, since technical developments have made it both faster and easier to imitate successful products. Third, the core competence is a basic source for developing new and unique products. Through global acquisitions, the firm may gain access to a more heterogeneous set of competencies. By recombining these in new and unique ways the firm can activate competence synergies that can be the source of new and innovative products.

- In order to increase competitiveness, develop new products and increase innovation, experience and insights must be shared about local market conditions, customers' preferences and reactions, as well as product characteristics and product developments. Setting up organizational communication arenas that support development and the diffusion of worldwide innovations is important for global competitiveness (Bartlett and Ghoshal, 1991). This may be obstructed by local units who want to keep their experience and insights to themselves because local knowledge is a source of power and local autonomy.
- To the extent that local subsidiaries consider synergies with each other, they seem to be biased, in the sense that it is mostly the exchange of existing products that is regarded as interesting. Competence and experience that are not directly related to the current product portfolio do not get much attention. There is often a lack of cross-national channels and arenas where insights and experience may be combined to develop innovative products.

Communication to reduce production costs

A frequently mentioned reason for globalization is the potential for reducing unit costs. Economies of scale, a result either of high fixed costs or learning curves, can be reaped by centralizing and combining activities across national subsidiaries and locating the activity where the costs are lowest. Globalization will therefore often imply a relocation of production activities. Economies of scale, and a need for standardization, may also lead to centralization of purchasing and other support functions, such as accounting, finance, human resources, IT, and research and development. Such activities will often be located at the corporate headquarters or the business units, although some activities usually also remains with the local subsidiaries. The location of the headquarters, however, may also change as a result of globalization, in order to reap benefits of being near the most important cluster of activities for the relevant industry.

Reducing production costs requires exchange of information about the costs of performing different activities at different locations. However, local subsidiaries may be tempted to resist a relocation or centralization of staff functions that reduce the level of activity in their own region. Activities with the purpose of increasing own production at the expense of other subsidiaries may arise, and local knowledge may be withheld in order to obstruct centralization of activities.

Corporate identity, coordination and management control

A global corporation must have coordinating mechanisms that maximize the firm's total values, preventing local suboptimization and integrating the values and cultures of the corporation. Such coordination is the purpose of the management system.

- A strategy that sets the basic goals and priorities must be developed and communicated. This requires communication processes through which subsidiaries provide information about local markets, resources and ambitions, and central units inform about what goals, actions and priorities it has been decided to implement. However, local subsidiaries may exercise influence in order to have the corporation set strategic goals and priorities that give priority to their local interests at the expense of others.
- Corporate vision and core values must be communicated in order to strengthen common identities, values and culture. Lean communication media such as reports, pamphlets, electronic networks, and telephones, appear to be less effective for such purposes than rich media like face-to-face interaction. This implies that values and culture should be implemented by organizational means that facilitate face-to-face interaction: overlapping leadership teams, for example. Attempts by headquarters to install a new corporate culture may be resisted in order to protect local values and identities, however. This may be especially important during globalization, since feelings of national identity may be evoked.
- Economic information must be transformed between the headquarters and the local subsidiaries in order to set goals for local units, allocate resources between different subsidiaries and functions, follow up economic achievements throughout the corporation, and provide information and handle requests related to non-routine decision making. However, economic information systems may in themselves invite local opportunism. First, local subsidiaries may fear that if they report results that are better than their budgets, the headquarter will 'punish' them by increasing the expected results for the next year. This is often referred to as 'Soviet mechanisms'. Second, a subsidiary may want to have some local slack in its resources in order to be able to serve its own local interests. Finally, local managers may expect that if they report a surplus

above what is budgeted, the headquarters will use this to subsidize other units.
- The headquarters should have the possibility to selectively intervene in the operations of local units in order to maximize the firm's total values. However, this may impose costs on the organization. One is the central decision makers' salary and the costs of providing information to support the system. Furthermore, individuals and units may have reasons to seek unproductive interventions, and they may spend resources trying to influence the decision maker to implement them (Milgrom and Roberts, 1992: 193). Furthermore, resources can be wasted if the system has an unclear specification of authority and responsibility. Decisions and implementations may be delayed, and time that could have been used for productive purposes is used in time-consuming decision processes.

Currently there is a dialogue going on between the research group and Nycomed about the topics to be studied in our future collaboration. One suggestion is to elaborate on the use of cross-functional teams, project groups and task forces, and analyze how such flexible organizations serve as new arenas of employee participation and horizontal problem solving within a global organization, and how they relate to hierarchical organizational forms and the established co-determination system.

This first analysis of Nycomed was performed within a top down-perspective. The problem was posed as one of identifying those communication processes and the the extent to which they influence the way company achieves its strategic ambitions, as seen from the headquarters in Oslo. In the next section we shall present a project that approach the issue more from a bottom up-point of view, focusing on how employee participation may be organized to improve business performance.

Developing new patterns of cooperation and participation: The case of Siemens AS

Siemens AS is a Norwegian subsidiary of the German multinational industrial giant Siemens AG. The Norwegian company is involved in several different business functions, such as production, sales and marketing, and the installation of electric equipment. The company is divided into divisions according to type of business.

Our collaboration with Siemens started in 1994. We were invited to participate in a conference that the firm had arranged. The topic was how employee participation and worker–management codetermination could be organized and activated so that it would improve business performance. The conference concluded that each business division should establish its own department council with equal representation from workers and management.

Department councils

Department councils have 50/50 representation drawn from workers and management. The chairmanship is rotated between the leader of the business division and a worker representative — often the head of the local union. As a general rule, the councils should not make decisions, but give advice on productivity, quality improvements, business strategies, budgets, staff changes, health, safety, and working environment. There is one council for each business division in the firm. Unless they are specifically invited to do otherwise, the councils deal only with issues that have relevance for their own division.

The department councils are coordinated and advised by the company council. This council performs the same function as the department council, but at the firm level.

This system of company and department councils is formalized in the main agreement between the Confederation of Norwegian Business and Industry (NHO) and the Norwegian Confederation of Trade Unions (LO).

Before we elaborate on the experiences with the department councils in Siemens, we present the model that influenced how we approached the issues.

Worker–management relations

The purpose of the research in Siemens was to support the company in finding out how the department councils could make employee participation consistent with improvements in business performance. However, the councils are only one element of worker influence. Figure 1 shows how employee participation takes place on different arenas, or put differently, in different systems of the firm.

In the bargaining system, employee participation is a right regulated by law and collective agreements. As they are formalized in the basic agreement between the national trades unions and employer federations, department councils are part of this system, which also includes all formal councils and negotiation arrangements that regulate workers' participation and worker–management codetermination. Certain business decisions have to be negotiated before they are implemented by management, especially issues of salary, working time and working conditions. In the bargaining system, norms of justice held by workers meet managerial norms of organizational rationality. Representatives of employers and workers hold complementary role sets, and they are defined as equal participants, negotiation parties, and, in some arenas, decision makers. In addition to the department councils, this system typically includes company councils, working environment committees, employee representation on company boards,

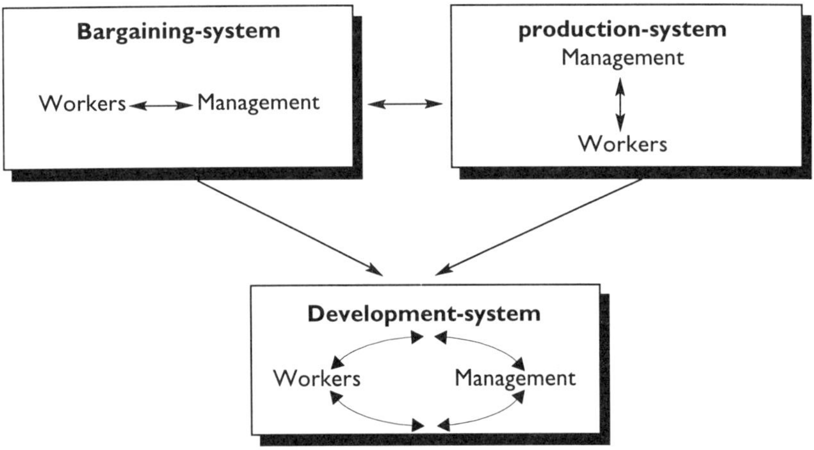

Figure 1: A system-approach to studies of participation

and institutions of wage negotiations.

The production system consists of all activities that are involved in a corporation's regular production activities. It is defined by the content of the work roles, and the relations between them. In this system, management are superior to employees. In contrast to the bargaining system, where workers and managers are defined as equal parties, there is a hierarchical relationship in the production system, although the degree of decentralization and autonomy may vary. The extent to which workers participate in decisions often depends on whether they possess informal power due to unique competence, tacit knowledge, and potential for informal collective action.

Companies may also be analyzed as development systems. Typically, such a system manifests itself in terms of project groups and task forces that take care of specific developmental needs. People are recruited to and leadership distributed within such groups according to competence, task experience, and reputation. The hierarchical structure of the production system is not necessarily reproduced. Independent of their formal position in the production hierarchy, all members may in principle initiate changes and develop ideas and recommendations through open communication, dialogue, and mutual trust.

According to the model in Figure 1, an important challenge for corporate development is to achieve consistency between the three systems. For example, the department council can, as part of the bargaining system, improve the functioning of the production system by providing information that is difficult to articulate in the production hierarchy. If such consistency is not achieved, however, the systems may obstruct each other. For example, if viewpoints that are articulated in the bargaining system are totally ignored or resisted by management, improvements may be difficult to implement throughout the production system.

Department councils in Siemens

Six of the department councils that Siemens established were located in Oslo. At the end of 1995 we delivered a study that described the leadership, member characteristics, and main tasks of these councils.

From the beginning, the heads of the business divisions were chairmen of the councils, while employee representatives handled secretarial functions. In some of the councils, employee representatives were elected by unions, while others had members that did not necessarily represent any union or profession. Some of the councils limited their agenda to questions of health and safety, while others had broader agendas.

Obviously, there was a lot of variation. At the time this first study was performed we summarized our observations by categorizing the councils into three groups:

- Department councils that limited their cooperation to questions of health, safety and and work environment, according to minimum requirements laid down by Norwegian laws and regulations.
- Department councils established as temporary arenas for learning and training on cooperation between management and employees. Such learning processes gradually broadened the organization and scope of the councils.
- Department councils that paid attention to any question raised, aiming at solving problems, improving organization of work and business development

At the end of 1995 three of the department councils were characterized by the first model, two of them by the second, and only one by the last model. This variation was more or less intended, as the councils were adapted to the specific problems and situation in each business division.

Our evaluation also showed that the department councils had led to improvements in the production system. Communication and mutual information between management and employees had improved. It appeared that managers got hold of information and became aware of problems that would have been hidden without these new channels of communication.

Today, after two years of experience with department councils, the challenges appear to be to improve the information flow between the councils and employees in general, to strengthen the councils as an efficient channel of communication and problem solving where anyone in the actual unit could influence the councils' agendas, and to include the councils as important parts of the decision making systems in the company. Our second study of the councils, conducted in the spring of 1997, concluded that:

- the councils have become arenas of cooperation, but this has not yet spilled over to employees and management in general;
- the councils are seldom involved in main questions of strategic importance, and their role in business development are mostly restricted to minor tasks.

The further development of the councils appears to depend to a great extent on managers' attitudes to codetermination and participation. In some of the departments, managers try to develop the councils into decision making and business development bodies, while the councils have only advisory functions in other departments.

Case analysis

During 1995 and 1996 we paid special attention to Siemens' department of installations in Oslo. This department provides electrical installations for hospital buildings, railroad tracks, tunnels, airports and similar heavy constructions. The department has 173 employees, most of them skilled electricians organized in a traditionally strong labor union. The leaders in the department are mostly engineers.

We were invited to take part in this division's bargaining council, department council, staff meetings, and management meetings, and we evaluated one of their main projects at an airport. We are now completing a survey among all leaders and employees on working conditions, education systems, level of employee participation, and so on.

Communications in this department have for a long time been described as conflict oriented. The purpose of involving researchers was to throw light on those barriers and thresholds that impeded constructive cooperation between management and employees. For example, we observed that most of the problem solving had a tendency to be approached by a conflict oriented negotiation style. The agreements and acts that regulate relations between the company and the union (NEKF) define a special salary system, connecting wages directly to productivity and effectiveness in each project. This system implies negotiations about wages, working time and working conditions during every project. The bargaining style that was appropriate for these negotiations, however, had become the dominating communication mode, even when it came to situations that could more appropriately have been handled by open dialogue and win–win approaches.

During the year in which we followed this business division, some changes did occur. First, the chairmanship of the department council started to rotate between managers and union representatives. Second, discussions and communication are now more free and open, and are characterized by a higher degree of mutual trust. This is illustrated by an event that happened in the summer of 1996. The union representatives proposed to freeze the wages in one of the installation projects for a period of three years, and have special agreements for each issue of working conditions in the actual project. The purpose was to gain some experience with how a different wage system would affect the performance of the project. Agreement was reached quickly, and signed by both parties after two meetings. The actual project has so far been more successful than others. Productivity is higher, conflicts more seldom arise, and the employees seem to be more satisfied with working conditions.

In the department council, management and employees have now agreed on a development process that includes the following issues:

- business culture and mutual roles;
- wage system and value destruction;
- competence and communication;
- project management and resource planning;
- allocation of responsibility and authority.

As this case study shows, worker–management codetermination may imply different logics of collaboration. In some negotiations, conflict-oriented bargaining is appropriate, while other issues may require a more open dialogue. An important part of a participatory approach to corporate development is to stimulate learning processes where the parties can differentiate their interactive style according to the tasks and arenas that are involved. In the language of organization theory, it appears to be fruitful to apply a contingency approach to the issue of employee participation.

Relations between Siemens and the research group

The researchers have stressed neutrality towards both employees and management in the processes described above. Our presence in meetings is concentrated on observations of communication, negotiations, and decision making, and the way in which the two parties contribute to these processes. The observations have been analyzed and reported to the actual councils, and to management and the union. These reports are then discussed and used as sources for improvements.

Sometimes we have been asked to comment on specific problems occurring in the meetings, such as special situations of blocked communication, the role of the different participants, and our overall views of the cooperative climate in the company. In one case we were included in negotiations as neutral advisers to both parties. Our recommendations were restricted to comments on how the parties can organize their problem solving processes and styles. We avoided suggesting specific outcomes.

In this project the research work has become an integral part of organization and business development, at the same time as these developments have influenced our research topics. Further experience is needed, however, to develop a clear understanding of the difference between the roles of researchers and consultants in this kind of project.

In the coming years we will continue the collaboration with Siemens on a broader scale. The intention is to use the results we have found so far to explore new forms of employee participation, and their impact on business development.

Experience from other Siemens plants

Before we conclude this presentation of the Siemens project we present some other experiences that this company have with employee participation.

Siemens has three plants in Norway, producing electronic and electrical devices. In one of these plants experiments on cooperation between management and employees were started in the early 1970s. Production groups were established, mostly based on ideas from the Tavistock school (Trist *et al.*, 1963), and studies on participation in Norwegian industry during the 1960s (Thorsrud and Emery, 1970).

In 1988 these production groups were developed by removing middle management in the plants. Now there are only three formal leaders. Authority is delegated to the production groups on matters such as production planning, job training, staffing, job rotation, quality, productivity, recruitment and disciplinary measures. Each group elects a 'contact person' for two years, who is responsible for communication between management and the group. Management takes care of strategic planning, budgets, and market relations. To coordinate activities between the different production groups, a communication group consisting of 'contact persons' from all production groups, union representatives and management has meetings about six times a year.

The role of the department council differs from the councils at other Siemens plants. Most of the tasks taken care of by department councils at other plants are solved in the production groups in the plant described above. The department council in the latter plant may pay more attention to the overall development and improvement of the business, as well as more general working conditions and decisions of importance to both parties.

Conclusions

The Department Councils in Siemens seem to be developing towards arenas which interact with all the three systems that we described in Figure 1.

- The councils involve representatives of organized labor and management, deal with questions of interest to both parties, and find solutions that the participants will introduce in formal negotiations on working conditions etc. In this way, the councils are part of the negotiation system.
- The councils provide information on and find solutions to such issues as productivity, cost reductions, and communication. A good dialogue in the department councils can both initiate and legitimize improvements in the production system.
- Discussions in the department council can provide new ideas that are taken over by project groups and task forces. When new solutions emerge in the development system, the department councils can play an important role in discussing the extent to which and how they can be implemented.

The case of department councils show the importance of achieving consistency between the different elements in the system modelled in Figure 1. The three systems should not be seen as isolated units. Bargaining, production and development are systems that are at work simultaneously, and they influence each other. The appropriate styles of cooperation, participation and problem solving vary between these systems, and an important aim of corporate development is to stimulate learning processes that enable the parties to apply one that fits the actual task and arena.

Participation and efficiency: A conceptual framework

We have presented two research projects within the 'Fafo/NHH' module. The first one identified communication processes that are important for corporate efficiency, while the second explored participation and worker–management codetermination. In this section we shall integrate these topics into a common conceptual framework.

Efficiency, value creation and value destruction

The efficiency perspective builds on the premise that the primary purpose of a firm is to create values. By this we mean that a firm must create a difference between the market value of a firm's output and the costs of a firm's input (Kay, 1993). To add value, the firm can either produce more of what the customers find it worthwhile to spend their money and/or time on, or it can reduce its costs, but not to a lower level than is necessary for its owners, managers, employees and suppliers to be willing to keep on supplying their resources to the firm. In Kay's words, added or created value is a measure of the loss which would result if a firm were to be broken up and the resources it uses deployed to other firms. A commercial organization which adds no value — whose output is worth no more than the value of its inputs in alternative uses — has no long-term rationale for its existence.

By defining value creation in this way, we build our approach on what is usually referred to as a stakeholder approach to organizations (Barney, 1997). Contrary to financial economics, which regards the purpose of the firm as being to create values that maximize the wealth of its owners, we regard the firm as a coalition of stakeholders who all have a valid claim on the values created, and who therefore also have a common interest in maximizing the values created by the firm, although they may have conflicting interests in how the values are distributed.

The other side of value creation is value destruction. Here, we shall say that value is wasted if some of the stakeholders supply an effort that makes actual output less than maximum output, for given inputs (see also Leibenstein, 1987). This may be caused by individuals, groups, or local subsidiaries who optimize their own local interests at the expense of the firm's global interests. For example, to secure their own position and level of activity, or to acquire resources to increase local spending and growth, a local manager may manipulate or withhold information from headquarters. This aspect of value wasting often arises as a result of what Milgrom and Roberts (1992) call influence activities. These are activities that attempt to influence others' decisions in a self-interested fashion, and often take the form of political activity or misrepresentation or distortion of information. Influence activities lead to influence costs, which consist of the time and effort that are taken away from productive purposes, and by the degradation of the quality of decisions.

Participation

Participation can normally be analyzed as a combination of decision making and implementation. The model in Figure 2 illustrates such a processual perspective on employee participation.

The essence of Figure 2 is to explain how different stages of a decision process may activate different modes of cooperation and participation from those involved. By participation, we refer to both formal and informal processes.

Any change process will start with identification of problems, and initiatives to solve them. Both employees and management may be the initiating agent. At this

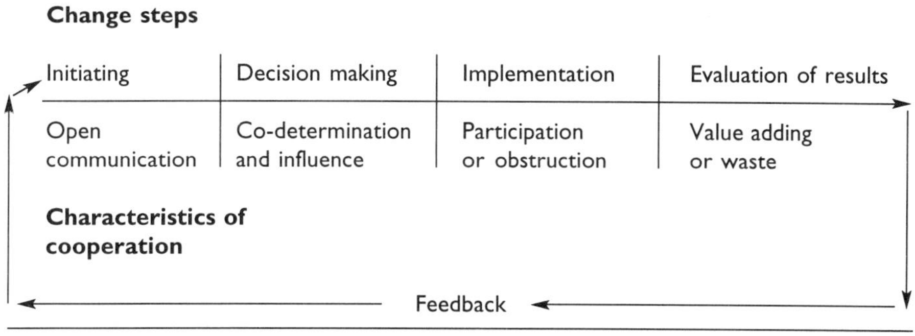

Figure 2: Participation as a decision- and implementation process (Falkum, 1997)

stage, a successful process demands open communication and exchange of assumptions and proposals.

The next step is to bring the initiatives to the point of decision making. Codetermination and influence may be measured as the extent of simultaneity in contributions from leaders and workers.

Studies have shown that successful implementation depends upon how specific the goals and ambitions of the actual changes are expressed and communicated, on the extent of agreement between the actors involved in the change, and on whether the amount and type of resources provided are sufficient and adequate to reach the goals of the decisions (van Meter and van Horn, 1975; Mayntz, 1975). Normally, employee codetermination in decision making will increase the probability of successful implementation. If the employees disagree with the chosen solutions, or do not believe in them, obstructing implementation may be a rational strategy.

In the final stage, results may be measured according to whether or not they increase the value of the company's outputs or, to put it differently, as value increase or value destruction. The important point to be made here is that to support implementation of a decision may be both productive or counterproductive, according to whether the decisions made were sound or not.

An integrating framework

One way to integrate the efficiency perspective with the participation approach is to combine the issue of value creation with different modes of employee participation. Figure 3 distinguishes between value creation and value destruction on the one hand, and two modes of participation. As far as the latter is concerned, we differentiate between interest articulation, which consists of activities where individuals, groups and departments promote their own interests and viewpoints,

	VALUE CREATION	VALUE DESTRUCTION
INTEREST ARTICULATION	Influence gains	Influence costs
RESISTANCE	Conflict synergies	Conflict costs

Figure 3: Employee participation and value creation

and resistance/obstruction, which are activities that are opposed to and try to reduce the influence of other stakeholders.

In Figure 3 interest articulation and resistance are combined with value creation and value destruction. The result is four concepts: influence gains, influence costs, conflict synergies, and conflict costs.

Influence gains refer to situations in which employe participation lead to added value. For example, as Freeman and Medoff (1985) have demonstrated, unions may represent information about employee preferences and possible labor process improvements that are not easily available to managers through other channels. Without this information, the firm may experience inefficiency in the form of staff turnover, low effort, poor quality, and technical breakdowns in production. Influence gains arise when the company has achieved consistency between the production, development and bargaining systems, and when participation and codetermination lead to efficient implementation.

Influence costs is the term that Milgrom and Roberts (1992) use to denote situations in which individuals and groups try to influence decision makers in a way that give priority to their own local interests, at the expense of the firm as a whole. However, such costs are not only due to the conscious efforts of local units to optimize own interests; they may also occur because the information and influence system is inefficiently organized, in the sense that it consumes a lot of time, distracts energy, and draws attention away from more productive activities. An example of this phenomenon is the existence of the 'Soviet mechanisms' that were referred to in the Nycomed study, which consist of budgetary processes where local units report and manipulate their figures according to what reactions they expect from corporate headquarters.

Conflict synergies refer to situations in which a dialogue between conflicting interests can result in new and creative solutions that give both parties a reward. These are situations where conflicts are transformed into win–win solutions. In the literature on bargaining this is referred to as integrative as opposed to distributive bargaining. In the latter there is a zero-sum situation which is not possible to transform into a win–win solution. Much recent management literature emphasizes such conflict synergies, as firms are increasingly acknowledged as being systems in which individuals and groups have opposing interests, identities, competencies, and sub-cultures. It is claimed to be a significant value adding potential in transforming such diversities into creativity. Resistance and obstruction may, if they are handled constructively, be a prerequisite for the emergence of creative solutions (Rognes, 1994).

Conflict costs arise when firms lack institutions, traditions and cultures that may lead conflicts into creative and integrative processes. The result is that different individuals and groups may use a lot of energy, not primarily to increase their own productivity, but rather to watch each others' actions, and in some cases even to try to destroy their success. To be on the winning side becomes more important than the actual results of the process.

Concluding remarks

This paper has investigated whether it is possible to integrate the efficiency perspective that traditionally has been developed and nurtured in business schools (NHH), and the employee participation approach that is used by, among others, FAFO.

We have given one example of each approach — a study of global communication in Nycomed, and an analysis of employee participation and codetermination in Siemens. The conclusion is that the two approaches can be synthesized in a conceptual framework that combines the articulation of employees' interest and resistance/obstruction on the one hand, with value adding and value destroying processes on the other. This framework will serve as an integrating baseline for future work by the 'FAFO/NHH' module within 'Enterprise Development 2000'.

Acknowledgements

We have received valuable comments and ideas from many researchers in the 'Enterprise Development 2000' program. We would especially like to thank Donatella De Paoli, Bjørn Gustavsen, Erik Jakobsen and Øyvind Pålshaugen. The final content of the paper is our own responsibility.

References

Barney, J. 1997. *Gaining and Sustaining Competitive Advantage*. Reading, Massachussets: Addison-Wesley.

Bartlett, C.A. and Ghoshal, S. 1991. *Managing Across Borders*. Boston, Mass.: Harvard Business School Press.

Colbjørnsen, T. 1997. *Strategi og organisasjonsutforming*. I: Tom Colbjørnsen (red.): Mangfold, medvirkning, tillit — ledelse og nye organisasjonsformer. AFFs Årbok. Oslo: Cappelen Akademisk Forlag.

Dølvik, J.E. and Steen, A.H. 1997. *Making Solidarity Work? The Norwegian Labour Market in Transition*. Oslo: Scandinavian University Press.

Falkum, E. 1996. *Styring, medbestemmelse og medvirkning*. Bedre Bedrift Nr. 1. 1997: FAFO. (Governance, codetermination and participation).

Gustavsen, B. 1997. "Enterprise development 2000: A program for distributive and interactive learning". This volume.

Hamel, G. and Prahalad, C.K. 1994. *Competing for the Future*. Boston, Mass.: Harvard Business School Press.

Mayntz, R. 1975. "External pressures and conflicts in the formation and implementation of environmental policy". Paper written for the Conference on Interorganizational Decision Making and Public Policy. Berlin: Unpublished.

Milgrom, P. and Roberts, J. 1992. *Economics, Organization and Management*. Englewood Cliffs: Prentice Hall.

Rognes, R. 1995. *Forhandlinger*. Oslo: TANO. (Negotiations.)

Scott, W. R. 1995. *Institutions and Organizations*. London: Sage Publications.

Stinchcombe, A. L. 1990. *Information and Organizations*. Berkeley: University of California Press.

Thorsrud, E. and Emery, F. 1970. *Mot en ny bedriftsorganisasjon*. Oslo: Tanum. (Towards a new business organization.)

Trist, E. *et al.* 1963. *Organizational Choice*. London: Tavistock Publications.

van Meter, D.S. and van Horn, C.E. 1975. "The policy implementation process". *Administration and Society* 6(4).

Chapter 4

Organization Development through Development Organization

Øyvind Pålshaugen

1. Introduction

The guiding principle underlying the research on enterprise development at the Work Research Institute (WRI) since the beginning of the nineteen-eighties may be formulated in the form of the following question:

How can work on enterprise development based on extensive participation, be organized and carried out most efficiently for the enterprise?

In practice this question implies the task of developing forms of work, methods, and techniques — in a word, instruments — for enterprise development. As is well known, there is a plethora of different methods and techniques in this area — at least on the supply side, which is dominated by the international consultancy industry in collaboration with university departments of organization and leadership, research institutes, and technical colleges. The research we perform belongs in this landscape, and extracts from it both practical and theoretical developments and innovations, as required (Pieper, 1988; Ropella, 1989; Davenport, 1993; Hammer and Champy, 1993; Tuckman, 1994; Ortmann, 1995).

During the last 10–15 years, however, the WRI has developed a niche for itself in this field, so to speak, with attendant niche products and a specialized competence. The basis for this specialized expertise is complex, but the following three aspects are especially important:

- The institutional framework of the Agreement on Enterprise Development between the parties in the labor market, the LO (the workers' federation) and the NHO (the employers' federation).
- The perspective derived from the theory of knowledge that constitutes the basis of our research presupposes, among other things, that research-based knowledge is not considered to have any priority over any other varieties of knowledge, but is a form of knowledge that is equivalent to these other

varieties; for instance the sort of knowledge which is found among the various actors in the enterprises.
- Our research projects have an action-research form which presupposes that the researchers themselves participate and perform tasks in the practical development efforts in the companies in question.

These three conditions have, in combination, given our research its distinctive character, which may be said to be its claim to distinction. We shall elaborate on this more thoroughly in the following three parts of this section. In the next section (2), we will present the principal purpose and the general strategy of the research module at WRI. In the final section (3), we will give an example of one of the six action research projects that our module comprises.

1.1 *The institutional framework*

The institutional linkage of our research projects to the Agreement on Enterprise Development has had two particularly important consequences. First, the agreement's stipulation that enterprise development work should be as inclusive as possible has led us to concentrate on developing the types of instruments that make it possible for *all employees* to contribute effectively to the development work in the enterprise (Pålshaugen, 1988, 1992; Øverland, 1994; Engelstad, 1995). Second, the Agreement on Enterprise Development rests on an understanding that the individual enterprise *itself* must determine what type of development measures people are to work with, as well as which results they want to achieve at the enterprise in question.

1.2 *The theory of knowledge perspective*

The theory of knowledge perspective upon which we base our research has emerged from a critical view of the relationship between language and reality (Gustavsen, 1988, 1990: 160 f.; Pålshaugen, 1992,1997). According to this view, all use of language implies that reality appears in the form of one *interpretation* or another. Language cannot *re-produce* reality directly as it 'is'. Language can only be used to represent or reproduce certain *interpretations* of reality. This also includes the scientific form of language use. Seen in this light, a scientific description is one type of interpretation which has no *logical* priority over other interpretations. This means that scientific knowledge must be regarded as one among many other forms of knowledge which constitutes no logical hierarchy. Instead, different forms of knowledge should be viewed as logically co-equal with one another.

Another way of expressing this is by saying that it is the *content* of the knowledge rather than its linguistic form that is the decisive point. This applies not only to the relationship between scientific knowledge and everyday knowledge. It also applies to the relationship between the various forms of knowledge that are found among different groups of actors in enterprises. This idea of the *equivalent* status of the different forms of knowledge constitutes the foundation of our research into the instruments of enterprise development. For these reasons, and because language is the most important instrument for the collective development of knowledge in companies — through dialogue and communication — then the development of *communicative* instruments is the centerpiece of our research efforts (Engelstad and Gustavsen, 1986; Pålshaugen, 1988; Engelstad, 1992).

1.3 *Action research*

The third element which gives our research its characteristic edge is that our research projects are carried out in the form of *action research*. This not only means that the researchers take an active role in the concrete processes of enterprise development that are the object of research: the main point is that it is this *active participation that is the very foundation of the researchers' efforts to generate new knowledge related to enterprise development*. Action research in companies has, as is known, been one of the dominating research strategies of the WRI since its inception at the beginning of the sixties (Thorsrud and Emery, 1969; Qvale, 1976; Engelstad, 1980; Gustavsen, 1990). Both the practical methods and theoretical justifications of action research have been refined over the course of the years. Even so, the basic question, which at the time resulted in the creation of action research, is still in effect the same: how can the results of research into working life find their practical application in — precisely — working life, in the enterprises? The question has not become less germane as time has passed. Recent research on the use of research shows that research results in the form of descriptive reports are no easier to implement today than they were previously (Beck and Bonss, 1989; Lindblom, 1990; Naustdalslid and Reitan, 1994).

The strategy of action research is to tackle the problem from an angle that is diametrically opposed to what has usually been the case. Instead of asking how new theoretical knowledge can lead to new practice, action research strategy is based on involvement in the development of *a new practice* (for instance, in the development of new methods for enterprise development) which, in turn, can create the foundation for the development of new theoretical knowledge. In this way we manage to make the practical problems the factors that guide theory development.

2. The principal purpose of action research in enterprise development

One basic point of departure, both historically and theoretically, for the action research at WRI, is a rather self-evident one: all work performed in an enterprise must be *organized* in some form or another. The words 'in some form or another' are just as important as the word 'organized'. The tradition within the theory of organization which constitutes the core of our research is known as *socio-technical* or *task-oriented* organization theory (Emery, 1959; Engelstad, 1972; Herbst, 1974; Gustavsen, 1979). One of the main points in task-oriented organization theory is precisely that there is *no single* organizational form which is the best solution to the problem of how work in an enterprise should be organized, which means in turn what the enterprise's work organization should look like. Instead, every individual enterprise has to find its *own local solution* to the problem of organization.

The general principle in task-oriented organization theory is thus that all organizing of work in an enterprise should take as its starting point the *nature of the work task* as well as the *technical* and *social (human)* resources which it has at its disposal to perform the task. In our elaboration of this task-oriented organization theory, we have drawn a line of demarcation between two main types of tasks with which all enterprises have to deal:

- *Operational tasks*, all of which are tasks that are carried out in the enterprise's chain of value creation. That is to say, all those tasks that — however indirectly — have to be carried out to get the enterprise's products out to the customer/market.
- Development tasks, all of which are tasks that must be carried out in order to *improve the conditions* for the tasks that have to be carried out in the chain of value creation, that is to say, the operational tasks.

While the operational tasks are those tasks that must be accomplished to achieve value creation, the development tasks are those tasks that must be accomplished to *increase* value creation. Common to both operational tasks and development tasks is that the work on them has to be *organized* — in one form or another. Also, the work on development tasks has to be organized in accordance with the nature of the tasks. Of course, the content of development tasks will vary at different times within one and the same enterprise as the enterprise development process advances. If the means and measures that research can contribute in support of the organization of the development work in the individual enterprise are to have any significance for enterprises in general, then these means and measures must be precisely such that they can *help enterprises find just the right ways to organize the development work, are in accordance with the nature of the development tasks.*

It is at this point that the following principle finds its application: that all those who are affected by the development work, as well as others who possess experience or knowledge of relevance to the content of the development work, must also be given the opportunity to participate in the work of finding the best solutions for the development tasks, a factor which, it goes without saying, is a precondition for achieving improved value-creation conditions, and, thus, increased value creation. The central issue for any enterprise that wants to find an optimum way to organize its work on development tasks will thus take the form of the following question: how can we organize the dialogues (between different groups of employees and managers in the enterprise) which are a necessary precondition for arriving at an effective way of *defining*, *organizing* and *accomplishing* work on the development tasks?

Of particular importance in our research in this field is the task-oriented theory of organization. To put it as briefly as possible, the main point is that by analyzing organizational aspects of both the operational tasks and the development tasks, it is possible, in cooperation with the enterprise, to determine *what type* of communicative instruments should be utilized so that management and employees *themselves* can succeed in finding the best way to define, organize and carry out the development tasks at *their* enterprise, based on local conditions.

In a manner corresponding to the way in which the total work on the operational tasks takes place in what is designated as the enterprise's *work organization*, we have designated the total organization of the work on the enterprise's development tasks as the *development organization*. We can thus summarize the research objectives in the following way: to carry out action research projects that contribute to the development of communicative instruments which can be utilized by a development organization in enterprises that are willing to invest in achieving increased value creation through extensive participation in development work. In the paragraphs that follow we shall discuss the most important subsidiary goals included within this general objective, and provide examples and specifications of them.

2.1 *How to create an effective development organization for the individual enterprise*

In what follows we shall present what we view as the most important challenges to research in this field. More precisely, this denotes two main themes:

- The question of the theoretical and the practical consequences of drawing a dividing line between operational tasks and development tasks.
- The question of which (set of) communicative instruments are the most appropriate for the individual enterprise in the different phases of a enterprise development process.

We shall devote one paragraph to each of these themes.

2.1.1 *Hypotheses on the importance of differentiating between operational tasks and development tasks*

One of our central research hypotheses is that an improved understanding of the relationship between *operational tasks* and *development tasks* will create a situation in which it will be easier to tackle these two types of tasks in practice, through the way in which work on them is organized. We shall therefore briefly present our perspective on the relationship between operational tasks and development tasks.

For the sake of simplicity, we shall designate all operations which have to be performed to get a product to the customer *operational tasks*. This applies whether or not the enterprise produces running series in a stable work organization or it produces single, tailor-made products by means of project organization. In both of these extreme instances the operational tasks are to be understood as those tasks which contribute to the creation of value because, however indirectly, they constitute a part of the production of a product on its way to the customer (the market).

In contrast to all such ordinary — both direct and indirect — operational tasks, which are all part and parcel of value creation, the *development tasks* may be defined as those tasks which, in principle, are based on the creation of *improved conditions for the execution of the operational tasks*. Seen from the point of view of value creation, work on the development tasks does *not* contribute to the value created during the product's progress through the enterprise's value chain. The whole point of working on development tasks lies in the promise of improving the *conditions for the enterprise's value creation*, which, *in the next round*, means that work on the operational tasks results in increased value creation.

Our hypothesis is that this perspective on the relationship between operational tasks and development tasks has three distinct advantages, which may have important practical consequences when it is elaborated in a more concrete form and incorporated into the enterprise's own practice:

- It helps to explain an important part of the reason why work on operational tasks as a rule has priority over, and often completely supplants, the work on development tasks.
- It shows that work on development tasks really ought to be viewed as a type of investment — or also as a type of development cost — and not as a type of operations cost.
- It is applicable as a point of departure for a 'tidying-up' of the apparently indistinct boundaries between project-organized *operational tasks* and project-organized *development tasks*. The distinction between the two types

of tasks shows clearly that it is the *relation* of the two types of task to the *chain of value creation*, not that they are project organized, that is the central point.

Moreover, and perhaps most importantly: taken together this indicates that the work on development tasks must be organized just as carefully and goal-consciously as the work on operational tasks. Combined with the goal of extensive participation in the development work, this brings us to the thesis that everybody should have the opportunity to take part in the *development* of the enterprise through the establishment and use of a development organization just as much as they participate in the *production* in the enterprise through the enterprise's *work organization*.

2.1.2 *Which communicative instruments ought to be included in the development organization in the different phases?*

WRI has employed a strategy in which cooperation between the enterprise and researchers is ordered into certain *phases*. The division into phases is partly decided upon on the basis of what the individual enterprise's development tasks consist of, and partly on the basis of the fact that the enterprise development process takes place as a cooperative R&D activity. One of the main points of this division into phases is that each separate phase will provide both the enterprise and us, the researchers, with the best possible opportunity of finding out how we can cooperate in a way that optimizes our combined needs to promote common interests, at the same time as each attends to their own special interests and needs (cf. Section 2.2 below). The division into phases which we have arrived at on the basis of action-research projects carried out during the nineties looks like this (Engelstad, 1996; Pålshaugen, 1992):

Phase 1: The link-up phase. This is the introductory phase. The researchers have been made aware of the enterprise's developmental needs and thoughts/plans about a development strategy. The enterprise has been informed about what the researchers can, in principle, offer in terms of providing assistance in organizing a development effort based on extensive participation. One precondition is that the dialogues between the enterprise and the researchers create the opportunity to demonstrate (or try out) one or several of the communicative instruments that will be employed when (or if) the R&D cooperation on enterprise development is started. The instrument that up to now has usually been tried out during the link-up phase is a *dialogue conference* in the form of a start conference for the enterprise in which as many as possible (preferably all) of the management personnel and employees take part.

The main objective behind such a start conference is to achieve a comprehensive discussion on what is, or should be, the enterprise's development tasks, and how work on these tasks ought to be organized, that is, how the enterprise can establish a development organization. Furthermore, by means of such a conference, the *enterprise* will be given a preliminary, but practical, experience of one of the central communicative instruments that research is able to offer, and the *researchers* are given a preliminary and practical experience of the enterprise's will and ability to make use of the development organization as an instrument in its development work.

Phase 2: The project development phase. This is the R&D project's main phase in which the enterprise development work is organized and carried out. The cooperation between the enterprise and researchers in this phase is based on the notion that the *enterprise* has the responsibility to draw up goals for the development work and attend to a goal-directed progression in the development work. The *researchers*, for their part, are responsible for helping the enterprise to find the instruments that are most appropriate, and to assist the enterprise in modifying and applying these instruments, as well as, not least, to improve them and develop new ones.

The dynamics in this cooperation may be briefly characterized as follows, as far as the production of research results is concerned: the *enterprise* places various challenges on the researchers' agenda by asking questions about how the enterprise's different development tasks may best be accomplished. The *researchers* generate their answers in the form of communicative instruments that they develop and help to implement through the enterprise's development organization in such a way that the enterprise's management and employees themselves can find the most suitable practical solutions to the development tasks.

It is a fact that the operating tasks, or the value-creation conditions, may, in principle, be improved throughout the *whole value chain* in the individual enterprise, not only at specific, selected points. The question of what can be done in the way of development and improvement at the individual stages in the value chain is not something management alone can answer; every single employee will be able help furnish solutions. This means that the question of what, at a given point in time, constitutes the enterprise's total development tasks cannot be answered unless management and all employees take part in a collective, purposefully organized dialogue aimed at finding out what these consist of.

The *dialogue conference* is one important example of the type of communicative instrument which aims at the generation of a total picture of the enterprise's total development tasks. The whole range of possible development tasks, from product development and productivity measures to matters concerning the working environment and security, will, as a rule, emerge through this instrument. In this way the enterprise's need for a development that involves products, the market, and customers will be linked to the employees' needs for development in areas related to training, the working environment, health, and security. Because the collective dialogue on these questions is organized in such a way that the discussions tie up both with each person's workplace and the enterprise's total situation, the discussions in a dialogue conference will lay the foundation for the management, in cooperation with the employees, to arrive at a total concept for an integrated, though differentiated, undertaking on prioritized development tasks through a locally adapted development organization.

- We have, by means of the action-research projects that we have carried out in conjunction with the Enterprise Development Agreement between the LO and NHO, and in connection with the Swedish LOM program as well as other, more independent, projects, gradually developed a considerable repertory of communicative instruments that can be utilized as a part of the enterprise's development organization (Engelstad, 1992, 1996; Gustavsen, 1992; Pålshaugen, 1988, 1992; Qvale, 1993).

One of the most central research task for us in the coming years will be:

- first, to further develop, improve, and renew this repertory of communicative instruments which can be included in the enterprise's development organization;
- second, to acquire greater and better knowledge about which communicative instruments should be included in the individual enterprise's development organization in such a way that the companies will be able to achieve the most effective results during the different phases of the enterprise development process, based on local conditions.

Phase 3: The institutionalization phase. This is the R&D project's concluding phase. What is hoped for is that the enterprise, during the first and second phases of the project, will have acquired so much theoretical knowledge and practical competence in the use of the development organization and communicative instruments in enterprise development work that, during the course of Phase 3, it will implement the measures (economic, organizational, and others) that should put the enterprise in a position to proceed with a continuous development effort based on extensive participation and without any special (or additional) support from the research community after the conclusion of Phase 3.

The actual results of the action research projects that we have conducted, both in Norway and Sweden, show that very few, if any, projects have managed to complete Phase 3, thus fulfilling the ambition mentioned above. On the other hand, more and more of the projects have advanced closer and closer towards that goal. It must be said, moreover, that an increasing number of enterprises in Norwegian (and Scandinavian) working life are showing considerable interest in committing themselves to continuous development work as a part of their business strategy. An extremely important research issue will thus be to find those instruments that can contribute to the realization of the enterprise's and the researchers' common ambitions for the third phase of the collaboration, as well as investigating what sorts of preconditions should be present. This is a topic, however, that extends beyond the relationship between research and the individual enterprise.

2.2 *Some general preconditions for R&D — projects on enterprise development*

In the next section we give a brief presentation of one of the R&D projects we are conducting in collaboration with a number of enterprises. First, however, it is worth mentioning one common feature of all these projects, related to the general conditions for the collaboration between the enterprises and our research group: our relation to each of the enterprises is regulated by a written contract, a 'Letter of Intent'. Each agreement has one general part, common to all of them, and one part which is specific for each agreement, relating to the particular development work to be done at each particular enterprise.

The general part of the Letter of Intent specifies a few preconditions which we regard as crucial for the collaboration to be an R&D project, with an emphasis on both *research* and development. The most obvious one simply states that the research team shall contribute to the resolution of practical development tasks in the course of the enterprise's development process. The reason behind this paragraph, of course, is to ensure that the collaboration aims at immediate results within the enterprise development process. The research team must, so to speak, contribute 'on line' to the practical progress of the enterprise development. If the research team thus fails in contributing 'on line', the outcome of the R&D project for the enterprise becomes too 'lean', and the collaboration comes to an end.

It is of great importance that our 'on line' contributions are not only a kind of *ad hoc* interventions, completely governed by the immediate needs of the various practical problems that occur all the time within any enterprise development process. If so, it would hardly be possible to develop either practical means or theoretical perspectives that are of some *general* interest; that is, practical and theoretical results which *other* enterprises and other researchers would also take some interest in gaining knowledge of, for their own purposes. For this reason, the agreement between any enterprise and our research team also contains a paragraph which states that the practical contributions from the research team to the enterprise development process, shall principally be of a kind that make such contributions bear upon some features that make them part of a broader development of new practical means and new theoretical perspectives. In practice, this means that the development strategy will be derived both from the institutional framework of 'ED 2000' as a national research program, as well as from the values of democratization of working life, which are the guiding principles for this research program. Thus, corresponding to the ultimate judgement of the enterprise as regards the *local* qualities of the 'on line' contribution from research, it is ultimately the responsibility of the research team to ensure that our *on line* contributions bear on the necessary *general* qualities.

Generally speaking, there will always be some tension between what we might call the local interests of the enterprises and the general interests of the researchside, when dealing with the questions of 'what to do' at any stage of the enterprise development process. This tension, which is always one of the constitutive elements in the dialogues that take place between the enterprise and the research team, may be regarded as a dimension of the fact that R&D projects of this kind are a collaboration between private and public institutions. From this point of view, the contribution from the *enterprise's* participating in ED 2000 is not only one of 'allowing' research to make the enterprises become even better at helping themselves. By participating in ED 2000, the enterprises also contribute to an enlargement of *the public sphere* concerned with the issue of working life development in ways that combine the demands for increased value creation and increased democratization.

Considerations like those above form part of the background to one more paragraph in the Letter of Intent: this concerns the written products from the R&D collaboration. In addition to the general clause governing the publication of any particular information about the enterprise, which can only be done with the express permission of the enterprise, this paragraph also states that *what* is written in research publications based on the R&D collaboration on enterprise development, will have to be decided upon by the research team themselves. One of the reasons behind this paragraph is that the scientific publications are aimed at a broader audience than each enterprise (or group of enterprises). More important is that the process of writing is a process of *developing and creating new knowledge* on the basis of practical experience; the process of writing is not one of simply documenting the practical development process (Pålshaugen, 1996). Documents may also be needed, of course. However, writing documents is by no means scientific writing. Thus, the Letter of Intent states that if there is a need to document the process, this is the responsibility of the enterprise.

3. The case of Kongsberg Aerospace

Kongsberg Aerospace is one of the enterprises within the Kongsberg Group ASA. The majority of the roughly 650 people who work at Aerospace are engineers. The main Aerospace products are anti-ship missiles, missile launchers, test and control systems for missiles, and some related mechanical production and maintenance services. The R&D collaboration between Aerospace and

WRI is anchored in the process of developing and producing a new type of missile. The main customer is the Norwegian Naval matériel command, which has signed a contract with Aerospace within a margin of expenditure amounting to 1.2 billion Norwegian Kroner (NOK) for the development of the new missiles in the year 2002. One of the conditions in this contract is that Aerospace shall enter into a partnership with a foreign enterprise, in order to make the new missile attractive to other countries as well.

This new missile will be considerably more advanced than any of the three previous generations of Penguin missiles which Aerospace had successfully produced and delivered to a number of military customers. Thus, the process of developing the new missile will confront Aerospace with new challenges, both technological and organizational. The number of persons working full time on developing and producing the missile will at the most be about 200.

To be able to cope with the many interrelated technological and organizational tasks, the enterprise management has decided to apply an integrated strategy for development and production of the missile, of a kind that is generally known as 'concurrent engineering' (Kusiak, 1993; Syan and Menon, 1994). As the general literature in this field states, 'concurrent engineering' designates an overall strategy in which both the product itself, the production processes, and the production equipment by which the product is to be manufactured, are developed *simultaneously*, in parallel.

By analyzing previous processes of development and production, both within the Kongsberg Group and from other enterprises, the management came to realize that concurrent engineering offers considerable potential for saving time and costs; the number of failures and deficiencies which usually will occur in the course of the development/construction process will be discovered at an earlier stage than by the ordinary, more sequential manner of development/production. Thereby, the costs generated from the work on correcting failures and deficiencies may be reduced considerably. In addition, there will be cost benefits, in that concurrent engineering means that the final product will require less expenditure, both in the production process and the maintenance activities. As a matter of fact, the Aerospace management regards a successful application of concurrent engineering as one of the preconditions for being able to fulfill their contract with the Norwegian Naval matériel command, and to be able to survive in the long run in the international market.

3.1 *The link-up phase at Kongsberg Aerospace*

The picture that was presented to us at the first encounter between Aerospace and WRI was more or less as described briefly above. The presentation was followed by an enquiry about what might be our contribution to a successful implementation of the concurrent engineering strategy. In general terms, our initial response to such requests is as follows: after having been told *what* the enterprise wants to obtain, we ask *how* they will obtain their goals; that is, by which means, working procedures and the like. After then having been informed about the enterprise's strategy or approach for implementation, our task will usually consist of pointing at any possible critical discrepancies between means and ends. Then we start to suggest adjustments and supplementary procedures that seem appropriate to us in that particular situation. This will serve as a point of departure for a dialogue with the enterprise on how to proceed further.

After having been exposed to our way of thinking, the Aerospace management wanted us to conduct about one dozen interviews/conversations with managers and employees at different levels within the New anti-Ship Missile (NSM) project. (NSM is the name of the project organization that was established within Aerospace with the responsibility to develop and produce the new missile.) These conversations should furnish us with the necessary minimum of information to give an initial feedback on the strategy for concurrent engineering. In order to make sense of this brief presentation of our R&D collaboration with Aerospace, we will have to present a minimum of the information and knowledge we ourselves gained through these conversations.

The work on developing and producing the missile is organized in one particular *project*, which has its main base within one of the six cost centers into which Aerospace is divided. The line organization within the cost center has the responsibility of supplying the NSM project with the required personnel resources, in addition to having the responsibility for the technical quality of the NSM project. The responsibility for progress, economy and technical solutions, of course, resides with the management of the NSM project.

This type of organization is quite common in enterprises whose products require different, high level competencies. The enterprise then needs highly skilled personnel from a variety of technical professions. To maintain the skill levels required, the personnel are usually organized in professional milieus located in different cost centers, from which the various product projects may 'gather' their personnel resources. This way of organizing thus serves to improve technical skills. Of course, it will also serve to create tensions and potential conflicts between the line organization and the project organization.

The NSM project is organized in a number of sub-projects, in a way which leaves each one of them with the main responsibility for each one of the main components of the missile. In other words, one sub-project takes care of the airframe, one the seeker, one the launcher, one the electronics, and so on. There is also one sub-project for the simulation system, since the whole missile, with all its components, is going to be simulated in a computer system, including a simulated launching, before any of the components are actually introduced in the production process.

All of those sub-projects involve a number of people with highly specialized skills within those particular professional competencies which are crucial for the development of each of the main components of the missile. In the end, of course, all the components will be put together to form the actual missile. The interface between the work that is done on each of those components thus becomes critical, and the concrete ways to cope with such critical interfaces becomes of crucial importance. From this it is easy to see that the cross-disciplinary communication between people who possess different technical skills and who have to interface with one another is of great importance.

General examples of such interfaces that have to be coped with are the relationship between the weight of the missile and its external dimensions; the relationship between the guidance system and the airframe, and so on. However, in addition to the interfaces between the main components, there are also a large number of critical interfaces between the different 'jobs' that go to make up each of the main components. Due to the high technical complexity of each one of those processes of development, the enterprise has to draft what it calls *process descriptions*, which give a schematic description of the way in which the various steps in the process of developing the particular components have to proceed, in order to ensure that the various parts of the missile are in accordance with the very stringent customer requirements. These process descriptions are transformed into an overwhelming number of technical specifications.

Last but not least, it is necessary to establish a superior *system process*, the purpose of which is to describe in a quite general way those procedures that have to be followed in order to be able to cope with the *totality* of internal dependencies and interfaces that exists in relation to the missile considered as a whole, that is, as a single product. As a matter of fact, in the initial phase of developing the missile, this means that the NSM project is performing a development task with an inherent structure of almost total reversibility. The relevance of the concept of concurrent engineering thus seems quite self evident.

Those conversations we had, however, revealed that the concept of concurrent engineering was not all that self evident. Or, to put the matter more precisely: the general intentions behind the concepts were more or less accepted, but not the way the Aerospace management had so far tried to realize those intentions by launching a separate 'concurrent engineering project' within the NSM project (abbreviated CE; IPF in Norwegian). The CE project should consist of two main parts. One part should furnish the NSM project with an IT system to make it technically possible to practice concurrent engineering. The other part concentrated on the working methods which had to be applied by the people working on the different parts of the NSM project, if the strategy of concurrent engineering were to be put into practice in the process of developing and producing the missile. It is, of course, this second part of the CE project to which WRI was invited to contribute.

Our first feedback to the NSM management was an appreciation of their intentions with the CE project, followed by a critical assessment of their efforts to put the project into practice. Briefly summarized, our critical points was: the practical commitment to the CE project from the NSM management was not sufficient; the resources they had devoted to the CE project seemed too close to or even less than the critical minimum; the implementation strategy seemed more like the issuing of a directive, rather than a participative one; and the timing was not too good. The last point was related to the contextual conditions. For some political reasons, outside the command of both parties, the missile contract between Aerospace and Naval Defense was at that time (autumn 1995/spring 1996) only of a preliminary nature. As a consequence, the intensity of the activities within the sub-projects varied somewhat, which, of course, affected the conditions for concurrence, in both meanings of the word.

In reply to our criticisms, WRI was invited to a two-day seminar together with the heads of the CE project, parts of the NSM management, representatives of a consultancy enterprise (ISI) assisting Aerospace on the IT part of the CE project, and a group from The Norwegian University of Science and Technology (NUST) in Trondheim. At this seminar, which involved about 16 participants, the overall strategy of the CE project was discussed in depth, in order to generate proposals for a common work schedule for the project, which included the supplementary contributions from the three external institutions: NUST, ISI and WRI.

On this occasion we launched the idea of making use of a dialogue conference for all the members of the NSM project, with the dual purpose of, so to speak, reaffirming or relaunching the CE project, and at the same time working within NSM on concrete discussions about how to implement the CE project in the best way. We were asked to work out a preliminary draft for a dialogue conference. This draft then became subject to further definition and revision throughout a series of meetings with key personnel, both from the management and the employees within the NSM project and from the line organization. We also involved people from NUST and ISI in this planning process, in order to be able to cope adequately with the intricate technological-organizational questions we faced in our effort to create a dialogue conference that was tailor-made to the particular situation at Aerospace.

The dialogue conference was held over two days in mid-March 1996, under our direction. It involved 91 participants from Aerospace. The main purpose of the conference was to improve the ability of those working within the NSM project to create and operate working methods of the kind demanded by concurrent engineering. In accordance with our basic concept of the relationship between development (or improvement) tasks and operational tasks, the dialogue conference was organized in a way that was intended to give each participant a maximum opportunity to contribute concretely to the discussion on how the strategy of concurrent engineering ought to be realized in practice, from the point of view of the participants' own heterogeneous, particular tasks in the process of developing and producing the missile.

A dialogue conference of this kind, in which almost everybody from the organizational unit in question participates, creates a new kind of *public sphere* within the enterprise. This public sphere, which is constituted by participants from all departments and levels of the organization, gives visibility to the total number of problems and possibilities for improvements in a way that is both more comprehensive and more impressive than the way they are experienced — or viewed — in the daily work situation. At first, when dominated by the mapping of all the problems, this public sphere may be experienced as a somewhat unpleasant method of provoking new ways of thinking. The main and lasting effect, however, usually is the constructive discussion that arises out of the initial impression of everything being reduced to ashes. In this respect, the public sphere created within an enterprise by a dialogue conference is, so to speak, an effort aimed at getting rid of the conventional dialectic of enlightenment.

Strange (or familiar) as it might be, this logic also worked at Aerospace. The majority of participants considered the dialogue conference an important contribution to a better understanding among the rank and file of the need to apply concurrent engineering working methods in the NSM project. In addition, the participants produced a number of proposals for action that might serve as a base for decisions to be made at different levels within the NSM project, after the conference. This material, together with a report drafted out by NUST on critical features of the NSM project, formed the basis for a grouping of the NSM project into 8 main areas for improvement. Four of these concentrated mainly on technical matters. The three points calling for further contributions from WRI concerned the creation of the work with the system process, working methods in cross-disciplinary work teams, and the strategy for a coherent system of information/communication within the different levels of the NSM project, as well as in the NSM organization as a whole.

Planning, running and following up on the dialogue conference was our main 'on line' contribution to Aerospace in the first phase of our collaboration. *They* were sufficiently satisfied with the outcome of our contribution, *we* were sufficiently satisfied with their willingness to invest in the use of communicative means on broad participation to form and improve their development organization. Against this background, Aerospace and WRI signed a Letter of Intent of the kind described at the beginning of this section. Thus, the signing of this agreement symbolizes a successful completion of Phase 1: the R&D project is linked up to the enterprise.

3.2 *The project development phase at Kongsberg Aerospace*

The contract between Aerospace and the Norwegian Naval matériel command was finalized in December 1996. The contract contained milestones along the missile development path that did not leave much room for hesitation. The 'operational' tasks of developing the components of the missile therefore very soon supplanted the tasks of *improving* the conditions for the design work on the missile components. That this would be the situation when the contract was finalized had been foreseen both by Aerospace and ourselves. However, the situation was complicated by some shifting of personnel within the management group. These two circumstances meant that it became more difficult for the staff running the CE project to implement the concurrent engineering strategy.

To cut a long story short, these circumstances continued in various ways to be an obstacle to any effective 'on line' contributions *directly* into the NSM project by us. We collaborated continuously with the people responsible for running the CE project, by a mix of partly discussing the overall CE strategy and its particular, concrete items, and partly discussing how to obtain a more effective realization of the CE strategy within the NSM project under the prevailing conditions.

The last mentioned type of discussions certainly were to some extent of an *ad hoc* character, and thus could not easily be reconciled with our research strategy aimed at developing *general* communicative means of enterprise development. However, this kind of *ad hoc* contribution is also one of our guiding principles in achieving our ambition. In addition, there will probably never be any R&D collaboration on enterprise development which is so streamlined or straightforward that the need for *ad hoc* contributions totally disappears. If that were so, then we would no longer be dealing with R&D; the 'output' would be 'industrialized'. Thus, the ability to make *ad hoc* interventions remains an important skill for an action research team. The point is that such *ad hoc* interventions should not cause the research team to 'go native', that is, to consider that the problem horizon of the private enterprise supplants the complementary horizon of the public research task.

In our case, the case of Aerospace, we worked steadily to come to a position where it would be suitable to directly combine the immediate development tasks of Aerospace with our research tasks of developing the communicative strategy of enterprise development. The next significant opportunity arose gradually during the spring of 1997. During this period, the management group of the NSM project was completed and became better consolidated, and both the strong and the weak aspects of the running of the NSM project became more easily identifiable, as the work of the sub-projects proceeded along the critical path, matching the milestones.

At this stage it became increasingly obvious that the NSM project would benefit from a more comprehensive process of consolidation, comprising all of the NSM project management, the line organization management, and a number of other key personnel. We therefore suggested another two-day dialogue conference for these groups of people, about 30 in number. The process of planning this conference differed from the planning of the previous one in two respects: first, the content or the issues to be worked with at the conference were very carefully linked to the specific tasks of progress that the NSM project was facing at that time. This was now possible, to a greater extent than in the previous dialogue conference, since the contract with the Norwegian Naval matériel command was now no longer preliminary in nature. Second, we worked even more carefully to anchor the responsibility both for the content of the conference and for the follow up after the conference with the NSM management. This was done to avoid those who ran the CE project becoming burdened with the task of following up in a way which exceeded their sphere of authority.

In sum, these two differences made a big difference in regard to the quality and effect of this second conference compared to the first one. The most significant general problem at that time was the phase adjustments between the different sub-projects within the NSM project. To cope with this problem, reallocation of personnel resources among the sub-projects seemed inevitable. Such reallocations might affect the ability of the different sub-projects to reach their milestones on time. Operating both with internal milestones, and external ones related directly to the contract with the customer, the whole problem also was partly one of definition: were some sub-projects too 'far behind', or were some other sub-projects too 'far ahead'? As will be understood, a situation like this may be an occasion for a great number of tensions, which are not to be cleared only by means of technical facts. The 'real situation' has to be constructed, by means of both technical facts and appropriate forms of dialogue.

Once again, to cut a (new) long story short, this second dialogue conference started with the 'political' question: to what extent is the NSM project in a state of phase displacement? By the same token, we also asked what contribution each sub-project could offer to bring the NSM project into line. The next issue was a related one, though framed and treated in a more strictly technical way: how to organize the work by establishing the so called *baseline* within a specified time limit? (A baseline is, metaphorically speaking, a 'snapshot' of the overall technical 'outlook' of the 'missile in progress' at a certain point in time.) Throughout the conference those issues were dealt with from somewhat different angles, in a steadily higher degree of concreteness. In other words, the discussions were organized in ways that also gave the participants an opportunity to gain a more *concrete* understanding of the relevance of the strategy of concurrent engineering.

In the end, our efforts to design and run this conference turned out to be a remarkable 'on line' contribution to the development process at Aerospace. The participants were unanimous in their evaluation of the usefulness and the outcome of the dialogue conference, which served both to get some piece of work done, and to consolidate the platform for the further work within NSM. From a research point of view, too, the actual state of the development work at Aerospace has given us a number of opportunities to reflect upon and generate new knowledge about how to work with development organizations. This knowledge, however, will have to be part of a new, and perhaps not too short, story.

Note

This article builds on the research plan and preliminary results from a set of R&D projects in industry, conducted under the auspices of the 'Enterprise Development 2000' Program (ED 2000). The research is conducted jointly by a team of researchers at the Work Research Institute: Øyvind Pålshaugen, Per H. Engelstad, Lars Mikkelsen and Erik Øverland, who took responsibility for the first two years of the project. Thoraf Qvale, Beate Karlsen, Henrik Finsrud, Karin Aslaksen, Øystein Fossen and John Bakke joined the team in December 1996 as a result of an expansion of the Institute's module within ED 2000. The present article thus reflects the work of the initial team of four.

Appendix: Other enterprises collaborating with WRI on R&D projects

Vestfold Bil and Karosseri (VBK)
VBK is a producer of bodies and special fittings for utility vehicles like ambulances, buses, trams, trains and ferries. The company employs about 140 personnel. VBK is located in Horten and exports 34% of its production. The R&D collaboration between VBK and WRI is aimed at a reorganization of the various support functions, to make them directly integrated into the value chain of the four profit centers within VBK.

OCEANOR
OCEANOR produces and delivers advanced technological instruments and services for use in oceanography. About 70 people are employed there. OCEANOR is located in Trondheim, and exports 50% of its production. The R&D collaboration between OCEANOR and WRI is aimed at implementing a quality system by means of participative methods, in order to prepare the ground for a more complete, systematic work on continuous improvements.

OSRAM ASA
OSRAM is a producer of lamps for the retail store market, in addition to a number of lamps for special needs. About 110 persons are employed. OSRAM is located in Drammen, and exports 10% of its production. The R&D collaboration between OSRAM and WRI is aimed at combined, general improvements in productivity, work environment and the cooperation between parties within the enterprise.

AS Elektrokontakt (ELKO)
ELKO is a producer of electrical fittings such as cables, switches and the like. About 250 persons

are employed. ELKO is located at two sites: Sandvika (just outside Oslo) and Åmot (north of Drammen), and exports 10% of production. The R&D collaboration between ELKO and WRI is aimed at establishing development groups to work on overall improvements to the benefit both of the conditions for value creation and of the working conditions within ELKO.

Norsk Hammerverk AS (NoHa)
NoHa produces various products for fireworks (pyrotechnics) and the like. About 150 person are employed. NoHa is located in Stavanger, and exports 70% of its production. The R&D collaboration with NoHa is undertaken in co-operation between WRI and Rogaland Research Center in Stavanger, and the project is focused on productivity and work environment issues.

References

Beck, U. and Bonss, W. 1989. "Weder Sozialtechnologie noch Aufklärung?". In *Analysen zur Vervendung sozialwissenschaftliches Wissens*. Frankfurt am Main: Suhrkamp.

Davenport, Thomas H. 1993. *Process Innovation — Reengineering Work through Information Technology*. Boston, Massachusetts: Harvard Business School Press.

Emery, F.E. 1959. *Characteristics of Socio-technical Systems*. London: Tavistock.

Engelstad, P.H. 1972. "Socio-technical approach to problems of process control". In L. E. Davis and J. C. Taylor (eds), *Design of Jobs*. Harmondsworth: Penguin.

Engelstad, P.H. 1992. "'Dialogkonferensen — en metod för forskningsstödd verksamhetsutveckling". In Jan Holmer and Bent Starrin (eds), *Deltagarorienterad forskning*. Stockholm: Studentlitteratur, 11–30.

Engelstad, P.H. 1995. "Fra dialogkonferanse til utviklingsorganisasjon". In O. Eikeland and H. Finsrud (eds), *Research in Action*. Oslo: WRI.

Engelstad P.H. 1996. "The development organization as communicative instrumentation: Experiences from the Karlstad Program", in S. Toulmin and B. Gustavsen (eds), *Beyond Theory — Changing Organizations Through Participation*. Amsterdam/Philadelphia: John Benjamins Publishing Company.

Engelstad, P.H. and Gustavsen, B. 1986. "The design of conferences and the evolving role of democratic dialogue in changing working life". *Human Relations* 39(2).

Engelstad, P.H. and Gustavsen, B. 1990. "Creating systems by dialogue — an emergement trend in organization development projects". In: *Organizational Competence in System Development*. Sweden: Studentlitteratur, 175–189.

Gustavsen, B. 1979. *Bedriftsorganisasjon — alternative modeller*. Oslo: Tanum–Norli.

Gustavsen, B. 1988. *Noen utviklingstrekk i aksjonsforskningen. Sociologisk Årbok*. Oslo: UiO.

Gustavsen, B. 1990. *Strategier for utvikling i arbeidslivet*. Oslo: TANO.

Gustavssen, B. 1992. *Dialogue and Development. Theory of Communication, Action Research and the Restructing of Working Life*. Assen: van Gorcum.

Hammer, Michael and Champy, James. 1993. *Reengineering the Corporation*. London.

Herbst, P.G. 1974. *Socio-technical Design*. London: Tavistock Publications.

Kusiak, A. (ed). 1993. *Concurrent Engineering, Automation, Tools and Techniques*. New York: John Wiley and Sons.

Lindblom, C.E. 1990. *Inquiry and Change: The Troubled Attempt to Understand and Shape Society*. New York: Routledge.

Naustalslid, Jon and Reitan, Marit. 1994. *Om forskningens rolle i politikk og forvaltning*. TANO.

Ortman, G. 1995. *Formen der Produktion. Organisation und rekursivität*, Opladen: Westdeutscher Verlag, 1.

Pieper, Rüdinger. 1988. *Diskursive Organisationsentwicklung*. New York: Walter de Gruyter.

Pålshaugen, Ø. 1988. "Ein Norwegisches Aktionsforschungsprogram zur partizipativen Demokratie". In W. Fricke and W. Jager (eds), *Sozialwissenschaften und Industrielle Demokratie*. Bonn: Verlag Neue Gesellschaft. (English version Oslo: AFI, 1995).

Pålshaugen, Ø. 1992. *Som sagt, så gjort? — Språket som virkemiddel i organisasjonsutvkling og aksjonsforskning*. Oslo: Novus forlag.

Pålshaugen, Ø. 1996. "This is not the whole story: On the demand for scientific reports of action research". In S. Toulmin and B. Gustavsen (eds), *Beyond Theory — Changing Organizations Through Participation*. Amsterdam, Philadelphia: John Benjamins Publishing Company.

Pålshaugen, Ø. 1997. *Kritikk av den ene fornuft*. Oslo: Spartacus Forlag.

Qvale, T.U. 1976. "A Norwegian strategy for the democratization of industry". *Human Relations* 29(5): 453–469.

Qvale, T.U. 1993. "The case of SBA: A 5 year programme on participation, productivity and institutional change". In T. Kauppinen and M. Lektonen (eds), *National Action Research Programmes in the 1990's*. Helsinki: Ministry of Labour.

Ropella, W. 1989. *Synergie als strategisches Ziel der Unternehmung*. New York: Walter de Gruyter.

Syan, C. S. and Menon, U. (eds). 1994. *Concurrent Engineering. Concepts, Implementation and Practice*. New York: Chapman and Hall.

Tuckman, Alan. 1994. "Ideology, quality and TQM in making quality critical". In H. Willmott and A. Wilkinson (eds), *Studies in the Labour Process*. London: Routledge.

Øverland, Erik F. 1994. "Bedriftenes erfaringer med bransjeprogrammene. Resultater og effekter på bedriftsnivå". In J.I. Karlsen (ed), *Samlet analyse av bransjeprogram*. HF-B rapport.

Chapter 5

Developing the Fish Processing Industry

Bergur E. Ágústsson, Odd Jarl Borch, Lene Foss and Knut Senneseth

Introduction

This paper originates from a module made up of two groups: one group is located in Tromsø and is working with the industry which processes fish caught in the open seas; the other is located in Bodø and is developing a program within the fish farming industry. The two contributions are presented separately below.

Fisheries are of major importance in Norway. The country's long coastline, and its location facing vast, open seas, mean that the historical role played by fishing as a major source of food as well as an export business is more important than in practically all other countries. With a current export value around 20 billion NOK (approx US$ 3 billion), fish today belongs to the largest export businesses in the Norwegian economy. Activities having to do with fish and fishing are widely distributed along the coastline, the Northernmost parts not excepted. In this way, fishing is a major factor in the regional policies pursued in Norway, with a strong emphasis on keeping the whole country not only populated but in principle enjoying the same standards of life and living.

The fisheries have experienced ups and downs over the years and much of the fish processing industry has been locally resource based, but oriented towards international markets and the demands they impose, as well as the oportunities they provide. In fact, one of the most dramatic changes in the whole Norwegian economy in the last decade or so has been a turn to international marketing and product development. Fishing has traditionally been subject to unpredictable changes in the fish stocks in the open seas. The fish farming industry, with its much more controllable conditions, has grown rapidly, adding another basic element to the stabilization and internationalization of this business.

From traditionalism to modern market orientation — the transformation of the fish processing industry

The industry and its development needs

The module located in Tromsø consists of about six researchers, of which some are part time affiliated. The module is a joint research effort drawing on researchers from the Norwegian College of Fishery Science, Department of Economics and Management, The University of Tromsø, Department of Social Science, The Norwegian Institute of Fisheries and Aquaculture Ltd (part of the Norut Research Group) and Norut Social Science (also part of the Norut Group).

The research activity is based on different, but complementary, theoretical frameworks. First, the intra-organizational perspective, which focuses on the internal processes of the enterprise; second, the inter-organizational perspective, which focuses on network relationships; finally, the benchmarking perspective, which focuses on measurement and comparison of enterprise performance.

The participating companies are located along the coast, most of them in the northern part of Norway. Part of the coastal areas of Northern and Western Norway can be characterized as economically depressed areas with an income heavily dependent on a fluctuating natural resource; the fisheries. To add to the problems, the same areas have experienced an outflow of people, especially young people seeking education and alternative sources of work in the cities and the southern part of Norway. To add to the discomfort of the remaining population, more females than males move away from these areas. What to do about such a problem is something on which views differ. This leaves productivity, in a broad sense, as a feasible strategy. Productivity, as the term is understood here, includes faster and more efficient work along any given production chain, as well as the development of new products and markets. This, too, means that all the theoretical and practical problems involved in the issue of productivity come to the surface.

In the fish processing industry, labor is not the dominant input factor when measured as a share of costs. Raw materials are far more central, accounting for 50% to about 80% of total costs, depending on the products produced and the production process used. Labor costs come in the second place, being about 15–20% of the total costs. The use of capital and the costs of energy come third and fourth. Improved productivity will consequently demand approaches in which all factors are subject to simultaneous improvement.

If a company is to undergo simultaneous transformation along a number of dimensions, the demand for the insights, experiences, ideas and competence needed to master all these transformations grows accordingly. One way of fulfilling this rising demand is for enterprises to co-operate in the development process — an approach that leads into the network perspective.

Turning to the third perspective — benchmarking — it is of critical importance to find a best practice company or production process for comparison. The theory of benchmarking is very vague about how to pick this reference company, or group of companies. Obviously, most could be learned from comparison with companies that are better than one's own. But how is it possible, from the start of a benchmarking process, to decide which are the best companies or production processes, and which are the inferior ones?

In the research part of the program, the development of a fruitful productivity concept, of inter-enterprise relationships that can support a new productivity dynamics, as well as of a benchmarking function that links the Norwegian fish processing industry to major international developments will be the core issues.

Developing relationships to enterprises: reaching a common ground

In developing the program, one of the first tasks was to visit several fish processing firms and talk with the managers about benchmarking and what kind of benchmarking was desired. Most of the managers we spoke with had a great interest in benchmarking; however they did not have detailed knowledge about benchmarking or benchmarking processes. Despite the fact that the Norwegian fish processing plants are operating in a competitive environment, most managers were interested in internal benchmarking as a first step. The reason for this is probably that most of the information about processing, production mix and raw fish supply strategy is often treated as sensitive information, and there is a reluctance to share information with competitors. Most of the managers were also interested in performance benchmarking on aggregate level for ranking companies according to performance level. Such information can be an important contribution to motivating and convincing the organization that improvement is both possible and necessary.

In regard to internal benchmarking, there is a need for the establishment of standards of performance, given the technological level of the production process as well as analytical tools for achieving these goals, or performance levels. On this basis a productivity measure will be established for each company. The productivity program will be divided into two parts. The first part will be to help the companies to develop measures which are understandable, precise and capable of guiding improvement. Useful firm level productivity measures should be able to satisfy following criteria Balke (1983):

1. Provide a clear understanding of the nature and causes of observed changes in terms of measurements.
2. Trace the effects of observed changes on traditionally reported financial and operating results, i.e. cost, prices, profit and return on investment (ROI).

3. Provide a sound indication of operations adjustments required to improve the results.
4. Provide a reasonable estimate of the impact of prospective changes in operations on the measures employed and, in turn, on costs and profitability.

Second, when reliable measures are established, the focus will be turned onto factors that effect productivity. These factors are (Boyd and Fleming 1987):

1. Motivation; when people do not derive a feeling of accomplishment from their jobs, they are not likely to do much more than necessary. Motivation that offers positive stimulation is therefor necessary.
2. Communication is essential in increasing productivity — this should not be underestimated.
3 Training and education of the work force is necessary for the firm to be able to produce products at low cost and high quality.
4. Work environment; productivity can be improved by creating an enjoyable work environment. Management can help to make work more enjoyable by displaying a positive attitude toward employees and by allowing them to make decisions.
5. Quality control; productivity is inextricably related to quality control. A product must be reasonably useful or otherwise it is worthless.

Our productivity program is therefore based on quantitative and qualitative approaches, which, of course, depend on each other.

One area stands out as being of major interest to most companies in the industry. In the fish processing industry the wage levels are mostly based on individual performance, i.e. the number of kilos produced, percentage yield, and a quality component, measured as a number of defects. This system leads to great differences in wages between workers, and at the same time the physical strain leads to high level of absence among the workers.

On this basis, the companies presented in Table 1 have agreed to co-operate in a joint development program.

Table 1. Description of the companies in the Tromsø module.[1]

Company	Aarsæther Vardø	Bodø Fiskeind	SL-Fjord Gruppen	Herøy Filetfabrikk	Herøy Laks	Seafood Vardø	Vevelst. Fiskeind.	Havprodu kter	Nilsen H Nilsen	Båtsfjord bruket
Turnover Mill NOK	132,1	55,9	42,9	32,3		50,2	7,1	125,8	90,1	57,2
Employees	135	22	80	45	20	43	10	160	100	60
Processes	Freezing /salting cod fish	Freezing pelagig herring	Salmon production	Salmon production	Salmon production	Freezing shrimps	Salmon prod.	Freezing/ salting cod fish	Freezing/ salting cod fish	Freezing/ salting cod fish
Part of concern	Yes	No	No	Yes	Yes	Yes	No	Yes	No	No
Training of middel managers	No	No	No	Yes	Yes	Yes	No	Yes	Yes	Yes
TQM	No	No	No	No	No	Yes	No	No	No	No
Foreign workers	50%	30%	7%	12%	12%	10%	0	50%	30%	10%
Number of nationalities	6	3	5	6	6	3	1	15	7	3
Labour unionisation	80%	50-60%	75-80%	96%	96%	95%	70%	70-80%	95%	90%

1) In addition to these 10 firms, we originally had a large concern within aquaculture and fish processing. Due to a merge with another concern our work in this firm is postponed. The observations from this firm is included in this paper.

Establishing the development process

Finding suitable data concerning performance benchmarking can be a problem. We have at our disposal production and financial data from a representative sample of production companies in the Norwegian fish processing sector. The data cover freezing, salting, drying and other ways of processing fish, and have been collected annually since 1977. We will also construct a questionnaire in order to gather data for the benchmarking study that are not available in our database. By putting these data together we will be able to identify the best practice companies in the fish processing industry, which will be used as benchmarking partners in setting performance standards for other companies.

Not all firms have been equally explicit with regard to their expectations of the project. We have responded to specific demands and needs. In order to cover our theoretical concerns and research questions we have developed an interview guide which covers the following topics: demographic data, formal structure and information handling, quality management, competence, leadership, social networks and organizational culture. These topics are covered in all firms, and questions relating to these matters are put to employees at all levels. The interview guide is semi-structured. Some questions are fairly open, aimed at getting adequate descriptions from key informants. Other questions are more structured, e.g. those pertaining to social network schemes where respondents are asked to specify relations according to frequencies and content.

We have put considerable emphasis on communicating our findings to the companies through confidential reports and oral presentations. We also advise our firms on organizational issues.

The findings so far show that the majority of our firms do not use formal structures (planned meetings) to any great extent. Information is disseminated *ad hoc*, and it is often passed around too late. One case shows how the spreading of information and the building of formal structures are difficult when the firm is growing rapidly. Generally, it appears that the firms are not doing well when it comes to the dissemination of information from the top and middle/line management to employees at the operating core. Horizontal communication, between the different departments, also leaves much to be desired. Our advice to the firms has been to formalize relationships through regular meetings to improve coordination both vertically and horizontally. Furthermore, the use of newsletters may be enhanced.

Another finding, which applies, in particular, to the larger firms in our sample, is the low degree to which top leaders communicate face-to-face with employees at the operator level. This finding is supported by personal diaries of networking among top and middle management (see below) and by interviews with employees and managers. External pressures, notably in the form of insecure market positions and problems in the concern of which the firm is a part, seem to shift the attention of top leaders towards internal leadership tasks. Our advice is to stress the importance of symbolic leadership, and to establish routines for regular interaction so that the top leader is visible in every part of the organization.

One case shows how the geographical dispersion of individual units in a concern increases the need among employees for information about goals, market position and business strategy. Employees in this firm scored lower than average on developmental themes, and higher than average on operational ones. Our advice to the concern is therefore to integrate the units better with respect to future policies, plans and development projects.

The use of personal networking diaries among top and middle management enabled us to uncover attention 'overloads', questions and tasks among submanagers. Furthermore, the analysis showed that relations with the employees were based too much on written information, and that there was little interaction with employees at the operating core. These findings were supported by qualitative data from interviews. The leadership took our advice and decided to spend more time on face-to-face interactions with employees, to shorten the vertical span of the organization and to start a 'joint committee' of leaders and employees to work out a common approach to future policies and projects.

The network perspective

The concept of social networks, which has its roots in anthropology and sociology, refers to a pattern of relationships between actors. The essence of this perspective is to predict individual and organizational behavior from the social structure in which these actors are embedded. The concept can be used to study social relations within and between the environments of organizations (Powell and Smith-Doerr, 1994). In this project we have used network theory to uncover internal problems in the organizations, such as the spreading of information, degree of contact between departments and between different levels in the organizations, and the degree to which developmental projects are communicated. In our sample we have three firms whose efforts at collaboration have been admired in the industry for a long time.

We measure networks by asking our informants in the firms who they have most contact with in their job: (a) within the firm, (b) in the concern (if the firm is part of one), and (c) externally (with suppliers, customers, organizations, institutions, for example). For each category, up to 10 persons are described, with initials and positions. The informants then describe the content of these relations. The content is grouped into four categories: operations (orders, one way communication), operations (problem solving, two-way communication), financial matters (budgeting, wages etc.) and quality and personnel management matters. The relations are further specified according to whether they are face-to-face, via the telephone, post or fax, and whether they are informal or formal meetings. In addition to standard network dimensions such as size, range, density, etc., these measures give us information about the use of formal structure and the qualitative content in social relations between individual employees.

In using network analysis to trace the pattern of communication among people at all levels in the firms, we found that inter-enterprise collaboration is mainly between the top managers, who meet informally, and who also see the symbolic value of cooperation; as a way of signalling strength *vis à vis* the environment. The collaboration among middle managers was mainly about borrowing materials, and could have been improved in regard to purchasing. The collaboration has been most successful in joint efforts on the supply side and least successful forwards in regard to the market, in areas such as production development and marketing.

Business strategy challenges and organizational development in small export firms

Introduction

The target industry in this project is the rapidly growing Norwegian fish farming industry. During the 90s, there has been a revolutionary change in the market relationships of this industry. Earlier, the industry was characterized by small, regionally based companies with few links to the market. Today, the industry has become globally integrated, with more concentrated, large-scale production and distribution, including trading networks throughout the world. The Norwegian salmon industry is very dependent on these markets, exporting 95 % of its total output. With a growth rate of 15–30 % per year, the small and medium sized enterprises of this industry have experienced considerable growth pains, especially in respect of the management and motivation of the employees. In this study we present the strategic dispositions of the industry, its opportunities and pitfalls in the international markets, and the organizational development efforts within the program 'Enterprise Development 2000' to improve market adaptability, competitiveness and employer–employee relations.

The industry

From its early beginning in the late seventies, the aquaculture industry in Norway has grown in a fifteen year period to produce export sales of about 7 billion NOK

(US$ 1 billion). The return on equity has fluctuated, reaching an average of 10–12% in 1995. Among the reasons for this success are good natural conditions along the Norwegian coastline, a well-developed regional infrastructure, and a motivated set of local entrepreneurs with a great ability to learn from each others' experience as well as from the traditional fishery industry (Borch, 1994b). Also, the Norwegian government has followed suit, offering financial support and, not least, publicly funded research programs have given the industry a technological edge compared with its international competitors. About 95% of the production is exported, 60 % of this to the European market.

Traditionally, the industry was highly fragmented, consisting of about 800 independent companies, each with only one production plant. Today the number of companies has been reduced to approximately 350, some of them multinational companies with production plants in several countries. The total direct and indirect employment within the Norwegian salmon industry is about 14,000 jobs.

This increased global supply of food products has worked together with a concentration in the distribution system following the growth of large wholesaler/retailer chains and discount stores. International cooperation has led to the establishment of international wholesaler and retailer networks. These networks are continuously benchmarking their suppliers, knowing where to find the quality and amount needed at the lowest price. This development has dramatically increased the negotiating power of the buyers of salmon in the value chain. As a result, the salmon farmer is faced with demands for lower prices, while at the same time having to balance the level of quality in the product. The change in market demands and competition in the end-user markets travels rapidly through the value chain, reducing the response time, thus placing increasing strain on the ability of the organizations to adapt and the motivation of the work force.

The strategic development problems

For the Norwegian salmon industry, as well as the rest of the international food industry, the challenge lies in how to exploit the opportunities of global markets, and at the same time avoid the pitfalls of high intensity competition. The second important challenge to the firm is to motivate its employees and increase their competence to match the dynamics of the market. The strategic competitive measures of the industry have been to:

1. Increase production to achieve advantages of scale. After 1994, the growth rate of the salmon industry accelerated, from a growth of 8–10% to 30–40% per year for the most dynamic companies. Even though the prices have fallen, the industry has managed to earn money through increased productivity and technological improvements. The production per employee has increased by a factor of three in ten years.

2. The industry has increased its knowledge of how to differentiate the product for different market segments: for example, the color, fatness and technical quality of the salmon for the smoking industry.
3. There have been changes made within the organizational structure of the industry to achieve scale advantages in the administration as well. A few multinational companies have entered the industry through corporate acquisitions, and larger regional constellations have developed through horizontal integration. The smaller companies have formed strategic alliances and created joint ventures for negotiations with both suppliers and buyers.

The business development tasks

The strategic changes mentioned above have created a need for managerial attitudes that will adjust the internal resources of the firm to fit the competitive strategy. In particular, the owner-managers have to secure a high level of loyalty and working discipline from the production staff at each plant. In the entrepreneurial phase of the eighties and the first part of the nineties, direct supervision, and the strong ties between the owner-manager and the workforce resulting from daily, direct interaction, was a significant factor contributing to growth and profitability. Among other things, the working productivity increased by a factor of three during this period. One important challenge to this managerial regime has been the growth of the enterprises and increasing the control span of the manager, complicating the direct employer–employee relations of the smaller entrepreneurial firm.

The research team of the ED 2000 Bodø module started by mapping some of the main organizational challenges of the typical Norwegian fish farming enterprise in the following areas:

- The improvement of the internal technology of salmon farming to keep up with the quality level or to differentiate quality, while at the same time being competitive on price internationally.
- How to institutionalize the development of knowledge and diffusion within the company, through inter-organizational relations (a) to match the rapidly developing technology related to breeding, feed composition, feeding regimes and sea cage technology, and (b) to acquire market intelligence to allow optimum contractual decisions to be taken, taking into consideration the production lead time of about 15–20 months.
- The continuous motivation of the workingforce for business and productivity growth, and formalization of employer–employee relations, including health, safety and environmental systems (HSE).

The points of departure

The points of departure were extracted from three earlier studies: One of them was the study of business strategy within the Norwegian meat processing industry (Borch, 1994a). This study showed how the most successful companies had, in a few years, changed their strategy from a differentiation strategy to a combined differentiation *and* low cost strategy; alternatively, a differentiation focus strategy concentrating on fewer products. This study gave us input on the market challenges and the strategic alternatives for the salmon industry.

The second platform was found in a participant observation study of salmon export companies, where the point of the effort was to link business strategy theory with institutional models (Borch, 1994b). The conclusion of this study was that the companies within the industry were reluctant to surrender power to new joint ventures. The companies wanted to remain flexible, and it was difficult to build the necessary trust and obligations to form an effective, long-term, horizontal, cooperative relationship.

The third theoretical cornerstone of this module was the findings from research on the institutional links of small and medium sized firms to regional structures (Aghed and Senneseth, 1992; Johanisson and Senneseth, 1993). These studies focused on the risk that SMEs with a limited strategic apex and strong owner-manager dominance did not manage to break out of a well-established pattern of adaptation anchored in their local community ties. Strong bonds to family and friends, the feeling of local responsibility and local, experience-based learning worked together with loose links to national and international arenas to create limited motivation and knowledge for change in this category of organizations. For an industry facing new combined challenges of differentiation and price competition from competitors, even on other continents, one must renew its institutional foundation to build a higher technology level within its resource relations, and to create a system for continuous adjustment in links at an international level (Loveridge, 1993).

The setting

The development efforts and in-depth studies take place within a regional cluster of companies in the second-largest salmon producing county in Norway (and world-wide) — Nordland. In this cluster we found representatives of all the different strategic groups in the Norwegian salmon industry. Also, there is an institution for regional education similar to organizations found in each of the other salmon industry regions. The firm size varied from two to approximately 30 employees.

The following companies participate in the project:

Firm A (regional branch of the largest international salmon manufacturer)
Firm B (small, owner-manager company)
Firm C (small family-run company)
Firm D (small, family-run company)
Firm E (small, downstream integrated, family-run company)
Firm F (medium-sized, geographically diverged company)
Firm G (medium-sized, upstream integrated company)
Firm H (farmer-owned regional educational organization)

The process

The ambitious goals of action research — to increase competitive advantage in a turbulent business environment — created several challenges to the researchers. The special setting of the aquaculture industry, with a large number of small companies, was an extra challenge compared with other, more integrated industries with only one or a few larger companies (Borch & Arthur, 1995). Among the challenges in this action research project were:

a. How to legitimize the research project and gain access to the smaller farms with limited time for 'academic' discussions, and a sound skepticism toward researchers.
b. How to deal with the range of practical development problems of the different companies.
c. How to accumulate the appropriate practical knowledge to contribute in solving complex problems.
d. How to diffuse knowledge to the rest of the Norwegian industry.
e. The development of theoretical hypotheses from working on practical problems.
f. Verification of hypotheses, leading to more general theories that add to the scientific field.

How to legitimize the research project and gain access to the farms

The setting consists of small companies with a limited strategic apex. The challenge was to gain access to the companies. The researchers could benefit from good relations with the regional branch of the Norwegian Fish Farmers' Association (NFF), and the well established reputation of their Institute. Also, some of the managers were former students. One of these managers was elected President of the NFF during this period. Third, a reference group was established with members from the different firm categories.

In spite of these efforts, the researchers had problems in gaining access, especially to the smallest companies. A solution to this problem was an alliance with the farmers' own regional educational organization, working with formal adult education programs for industry employees. Through this organization and through the regional office of the Norwegian Fish Farming association, meeting places as well as arenas for the diffusion of knowledge were created. Finally, the efforts during the first research phase of doing applied research in the most critical areas of the industry gave the necessary 'functional substance' in the discussion with the farmers.

How to deal with the range of different practical problems within the different companies

This was done by visits and initial interviews with all the owner-managers and most of the employees of the companies. Certain common areas of interest were identified from these initial interviews. These areas were then refined into research problems and the research projects resulted from an expression of interest by both the individual firm where the research was conducted, and by the cluster of firms. To contribute to the internal development process of the firm, we started by distributing the results of the (market-oriented) applied research projects through reports and newsletters. We then arranged meetings with the management and the employees, in particular the employee (union) representatives. From this discussion we prepared an internal 'plenary' seminar. This seminar was used to discuss the strategic challenges of the firm from both a company and an employee perspective. The practical problem was how to build a quality system for the firm that included competence development, health, safety and environment questions, as well as routines for employer–employee relations (TQM with Internal Control Systems). The research team contributed actively to the implementation of the plan, serving as consultants during meetings.

How to accumulate the right practical knowledge to contribute in solving complex problems

To achieve the necessary competence, the research team was composed of researchers with long experience both within industry research and consultancy. Two of the four researchers in the research group had experience as board members in agricultural companies. One of the researchers had been working in a governmental institution evaluating the industry. Second, representatives of the companies were included in the research group in the areas where the research team had less experience, doing part of the data collection. Third, the researchers interviewed a large set of actors among both suppliers (feed production companies), and buyers (export wholesalers and the smoking industry). We were able to benefit from another research project within the institute, focusing on the biological aspects of the salmon product quality. This project was associated with the Company Development 2000 project. Through the theme based projects, applied knowledge was created as a basis for internal development, competitive strategy and development networks.

Future activities

The dialogue within the companies and the external analysis of the industry confirmed the importance of relating organizational development efforts in this industry to knowledge about technology, governance structure and the functional tasks throughout the value chain.

During the first phase, the research team developed a quality management development system that builds upon the plans for employment health, work environment and job security (HSE) already present in most companies. The next step for the researchers is to serve as change agents within the organizations building upon this TQM/HSE outline. The researchers join the company as discussion partners in relation to such matters as internal routines, they arrange staff meetings discuss knowledge improvement and HSE questions, and they contribute to courses arranged by the regional educational organization on TQM/HSE subjects. The main focus will be on (1) competence network development both at management and employee level, (2) developing arenas for dialogue between the management and the workforce representatives, and (3) systems for internal communication as to both strategic and operational questions. The institutionalization of routines in this respect may help the smaller companies cope with the extreme volume growth and technological change within the industry. Through adjustment of the development program to the characteristics of the typical fish farm, the goal is also to facilitate the diffusion of knowledge of organizational development processes to the rest of the industry.

References

Aghed, S. and Senneseth, K. 1992. "Entreprenörens signalement". In B. Johannisson (ed), *Entreprenörskap på svenska - Affärer och förnyelse*. Malmö: Ekonomiförlaget.

Andersen. B. 1995. *The Results of Benchmarking and a Benchmarking Process Model*. Ph.D. Dissertation. University of Trondheim. The Norwegian Institute of Technology.

Armenakis, A.A., Harris, S.G. and Mossholder, K.W. 1993. "Creating readiness for organizational change". *Human Relations* 46(6):681–703.

Axelsson, B. and Easton, G. 1992. *Industrial Networks. A New View of Reality*. London: Routledge.

Bjørnåvold, J., Foss, L., and Moldenæs, T. 1995. "Behandlet material fra intervjuer foretatt i oktober 1995 - Aarsæther Vardø AS". *Arbeidsnotat*, konfidensielt, Fiskeriforskning.

Blake, T.R. 1983. "Some applications of productivity analysis in the petroleum industry". *Int. J. Mgmt Sci.* 11(4): 329–341.

Borch, O.J. and Arthur, M. 1995. "Strategic networks among small firms: Implications for strategy research methodology". *Journal of Management Studies* 32(4): 419–441.

Borch, O.J. 1994a. *Markets in Transition. A Study of Business Strategy within the Norwegian Meat Processing Industry*. NF-Report 8/94, Nordland Research Institute, Bodø.

Borch, O.J. 1994b. "The process of relational contracting. Developing trust-based strategic alliances among small business enterprises". In P. Shrivastava, J. Dutton and A. Huff (eds), *Advances in Strategic Management*. Greenwich, Connecticut: JAI Press Inc.

Boyd, P. and Fleming, M.M.K. 1987. "Productivity: The controller's role". *Management Accounting* September.

Foss, L. 1996. "Tallmateriale basert på dagbokføringer hos ledelsen i Aarsæther Vardø AS". *Arbeidsnotat*, konfidensielt, Fiskeriforskning.

Foss, L. and Moldenæs, T. 1996. "Behandlet materiale fra intervjuer høsten 1995 - Seafood Vardø AS". *Arbeidsnotat*, konfidensielt, Fiskeriforskning.

Foss. L. and Nilsen, S.R. 1996. "Tilbakemelding på dagbokføring hos topp- og mellomledelsen i Seafood Vardø AS". *Arbeidsnotat*, konfidensielt, Fiskeriforskning.

Foss. L. and Strand. H.B. 1996. "Informasjon, kommunikasjon og kvalitetssikring - Frøya Fiskeindustri AS". *Arbeidsnotat*, konfidensielt, Fiskeriforskning.

Hamlin, J.L. 1978. *Development in Firm Level Productivity Measurement*. The American Productivity Center. 1978.

Johannisson, B. and Senneseth, K. 1993. "Paradoxes of entrepreneurship". In H. Klandt (ed), *Entrepreneurship and Business Development*. U.K.: Avebury.

Loveridge, R. 1993. "Socio-economics - A bridge too far?" *Human Relations* 46(9): 1029–1034.

Chapter 6

Integrating Engineering and Social Science Knowledge in Participative Enterprise Development

Morten Levin, Eirik Borgen, Reidar Gjersvik, Roger Klev, Ida Munkeby, Monica Rolfsen and Hans Jørgen Sæbø

Introduction

Enterprise development is a concept that has its origin in the social sciences. It has for a long time been a standard term in the vocabulary of joint management and union organizational developmental initiatives. Most of the important enterprise development activities are in actual practice not joint efforts. Thus, employees are basically given a reactive role in these change projects. It also seems fair to state that these fundamental change processes in companies have to a very great degree been the responsibility of engineers and economists. This paradox is important. Employees are basically reactive in the major enterprise developmental activities, while they are granted freedom to take responsibility for a proactive role in joint OD efforts. This is the position that underlies the agenda for the Trondheim module.

What if we could create a proactive employee engagement in the important change activities that are accomplished by engineers and economists? What if we could create a participative approach to organizational change that also involved major actors in engineering? This sets the agenda for our work in the Trondheim module of ED 2000. We find that in order to move towards such a goal, engineers and social scientists need to be able to work together, in the same way as employers and employees need to work together. The main thrust of the paper is to present and discuss possibilities and pitfalls in such a cooperation, where the main point is that the structure and process of cooperation determines the kind of knowledge that is produced. Further, we develop an epistemological reflection on the similarities between engineering research and action research approaches to knowledge construction.

Integrated enterprise developmental activities cannot be accomplished purely by mandating them in agreements or legislation. If integrated development is to become a reality it has to be based on new forms of cooperation. An understanding must be developed that create a potentially more efficient enterprise development based on a proactive engagement both within engineering circles and among social scientists, joining with local actors in the companies. This development at the same time grants increased control to local actors over their own working situation.

The research context

The Trondheim module is located on the premises of the Norwegian University of Science and Technology (NUST). A major element in the charter of NUST is the imperative to support the production of knowledge in the intellectual space created at the interface of science, engineering, the social sciences and humanities. Thus, the key problems related to multidisciplinary and transdisciplinary knowledge creation are essential for NUST. Researchers from the Department of Sociology and Political Science are engaged in the Trondheim module.

A non-profit research organization, the SINTEF Group, cooperates closely with the University departments and is located in close proximity to them on the campus. Almost every department at the School of Engineering has its own research branch, organized as a section of SINTEF. The Industrial Management Institute, which is one partner in the Trondheim module, has 130 researchers working in five departments. Researchers from three of these five departments participate in the Trondheim module. Those three departments represents a fairly wide range of competencies, spanning from industrial and production engineering, via health and safety, to organization and leadership. The following brief presentations highlight some of the individual characteristics of the departments.

The Department of Sociology and Political Science

The department is part of the Faculty of Social Sciences at NUST and it has a broad scope, spanning general sociology to political sciences. Work sociology has an important place in the teaching and research. Business development and organizational change has lately become a field of activity.

SINTEF–IFIM — Institute of Social Research in Industry

SINTEF–IFIM was founded in the late 1950s, and researchers from this institute had a central role in the Norwegian Industrial Democracy Project in the mid

1960s. The institute does research and development work aiming at giving increased understanding of and knowledge about working life conditions in both private and public sectors. The institute focuses, among other issues, on social and psychological conditions and the effects of industrial development. IFIM and the Department for Sociology and Political Science have a sustained activity in action research.

Safety and Reliability

This department carries out contract research for the petroleum sector, the onshore industry, transport, and the public sector in the fields of safety, reliability, maintenance and quality. Important research activities include the development of models, tools, databases and standards for the efficient treatment of safety and reliability matters. Maintenance engineering is important in the Trondheim module.

Production Engineering

This department does research and development work to improve manufacturing and logistics in companies. This is done through the development of integrated process-oriented solutions of all industrial processes. The concept of continuous improvements and total quality is a cornerstone in their work. Relevant areas for research in the Trondheim module are flow-oriented production, Just in Time supply, project management, logistics, production simulation, total quality management, and techniques for improving productivity.

This complex group, with different professional backgrounds and affiliations, contributes to the core research team in the Trondheim module. Some of the researchers have earlier experience with research groups that crossed the barriers between social science and engineering, some are experiencing it for the first time. Through teaming up with the companies in a joint effort to address enterprise developmental issues, we are gaining knowledge on integrated development. The concrete focus of our research is developed through interaction at company level.

The cooperation between engineering and social science

The Trondheim module has two parallel intellectual agendas. One objective has a conventional flavor as it is constructed around an action research approach to company development. In this respect, we are working in an area of knowledge that has been researched for many years. The core question, given this perspective, is to create a research agenda that both satisfies local needs and has intellectual content sufficient for shaping cutting edge research on company change. The communication processes between researchers and locals that are crucial in shaping this common ground for developmental activities are reasonably well understood (Elden and Levin, 1991; Gustavsen, 1992; Isaacs, 1993), although it is important to get a better understanding of how this common ground supports business development.

The second and most important objective is our research on how to integrate engineering and social science competence in company development. This process creates a new dimension in the relationship between researchers from the two very different disciplines. The challenge in creating a joint understanding between locals and researchers is still the same, but the new dimension is to build a bridge between engineering and social science knowledge (Levin, 1997). From being a dyad, the communication space is now a triad, each group modeling developmental activities according to different understandings and concepts. This communication involves the challenge of connecting the worlds of engineering and social science. Engineers are concerned with issues like productivity, benchmarking, production control systems, IT systems, quality control systems, etc., while the vocabulary of social scientists focuses on innovation, job content, participation, leadership, etc. Basically, both camps address the same issues, which, of course, are the organization and the production system. On the other hand, there is not necessarily a shared frame of reference lending a common language for communication. The shaping of this common understanding is the first step in a process to create an integrated approach to company development.

The epistemology in engineering and social science

The Trondheim module attempts to integrate engineering and social sciences, aiming at solving integrated company developmental issues. In doing so, the differences in scientific paradigms will easily emerge at the interface of the joint professional activity. To begin with, it would be a probable prediction that this kind of cooperation would be conflicting, as it might be expected that social science and engineering differ very much in their research paradigms. It would seems easy to embark from such a position, but in the following discussion we develop distinctions that run counter to this kind of argument.

An investigation into engineering epistemology reveals an interesting division of arguments. A conventional understanding of engineering work leads one to consider it as applied science, and hence closely linked to logical positivism. A scientist is a distant observer, practicing design based on the application of scientific knowledge. General laws direct attention in the design process. This kind of argument is only acceptable if one considers design as an activity performed by non-humans. Rather, we want to take on Bruno Latour's (1987) argument of technological development as a conscious human activity, shaping networks of actors and enrolling technological artifacts in a process of influencing technological design. Thus, design becomes an activity of creativity and network building in order to decide the potential outcomes of a technological artifact. In this case, the epistemology will be the epistemology of constructive social science (Berger and Luckmann, 1966; Guba and Lincoln, 1985). Claims of validity will be of a non-traditional form. If one does not focus on the conventional logical claims of validity and generalizability, one can substitute them by claims of workability (Greenwood and Levin, forthcoming). This claim represents a sort of intuitive and absolute for any engineering design of an artifact: it has no value if it does not work. The construction in itself must be able to solve the problems it is expected to solve.

Let us take a parallel trip into social science. Conventional social science has much in common with logical positivism. It applies a rule-based strategy to validity claims, relying on certain statistical measurements as a warrant for a trustworthy conclusion (Levin, 1995). Action research has a different slant on this, since validity claims are based on shaping social constructions that really work. Hence, the validity claims are based on experiencing solutions of actual problems that really work (Greenwood and Levin, forthcoming).

Our main argument is that engineering and action research approaches have much in common. The distinction between the two ontological, epistemological and methodological positions is not very different. This makes cooperation to some extent quite easy. It is not difficult to agree on basic issues of science and scientific work. The real challenge, however, is how to facilitate communication between the different professional positions. Every profession has a language of its own, with concepts, theories and methods of working that are specifically linked to one specific paradigm of science (Kuhn, 1962).

What we focus our attention on is the shaping of a shared space of understanding. Achieving this seems only possible through engaging in a continuous, open dialogue. We develop a common, new language through cooperative work on concrete company developmental issues. Experiencing the same reality and reflecting on issues from this experience makes it possible to bridge the gap between engineering and social science. The communication process is what creates the common understanding through the shaping of a common language (Wittgenstein, 1953; Johannesen, 1994; Monk,1990). The new language, having connotations to the same praxis, shapes the common space of understanding. This, of course, is what characterizes a shared language.

Shaping a shared intellectual space

We have constructed two parallel processes to create common experiences and to reflect on these issues. The primary organizational structure has been the establishment of joint company teams. In each of the companies we are involved in, a multiprofessional team has been put together, comprising both engineers and social scientists. The teams were constructed to bring to the company the necessary competence for what, initially, we expected to identify as important for the company. We are of course prepared to change the teams as the field of interest might change. To some extent each team has a member who has had a relationship with the actual company extending over several years. Each team bears joint responsibility for the work. We started with a mapping study to identify key developmental and research issues. This work resulted in a complex combination of joint understanding, new areas of understanding specific to each profession, and areas where new understanding is emerging.

The participating companies

Recruiting companies to the module was certainly a complex process of getting to know each other, figuring out the mutual benefit and just learning about how to create a mutual learning situation between people in the companies and the research team. We will deal with this process more in detail as we present the participating companies.

There was no single line followed in recruiting the companies to participate in the module. Most of the companies signed up because we had had a previous relationship. Some companies entered the module from the engineering side, having a vision that the engineering approaches they had been through had left open challenges to redesign their organization structure and leadership performance. These companies were recruited by the engineering members of the team. Other companies were recruited by social scientists on the basis of former relationships. A third recruiting mechanism was 'matchmaking' by the Research Council, which stated that a prerequisite for granting money to a technological developmental project was for the companies also to engage social scientists to take a look with them at the developmental issues. For the Trondheim module this attitude of the Research Council was very interesting, since it offered possibilities for the type of research we are interested in. Our final group of companies numbered five; one offshore construction yard, an office chair producer, an aluminum smelter, a subcontractor in the gas turbine industry, and a sausage factory.

In our presentation we will go more in detail in two of the cases, since that will illustrate the kind of work we are engaged in. The other three companies are just given a basic description.

The office chair producer

The factory is located a town situated some two hours drive from Trondheim. The company employs 250 workers, and is by far the dominant local employer. The construction and the design studios are located in the capital, Oslo. Almost half of the turnover derives from export. The financial results are quite satisfying, and the turnover has increased 15-fold over the last 10 years. The office chair producer has expanded constantly over the last years.

The company was recruited to the Trondheim module on the basis of a previous relationship with the Production Engineering department. The engineering research team had, over a number of years, restructured the production system from a conventional batch production system to a Kanban and JIT based system. This has been a successful transformation, but there are still some technological problems to be solved regarding the integration of information technology and the JIT production control system. A specific problem is caused by the daily variability in total orders.

Basically, the introduction of the new production control system was done without any particular focus on changing the organization. The organization's structure is the same as it was 10 years ago, when the company was much smaller. The existing organization is not particularly well adapted to the technological challenges caused by the new production system. It is fairly evident that there is no unified design of technology and organization. However, the new production system has led to improved competitiveness, both on domestic and international markets. The company has an extraordinary focus on an advanced design philosophy, the key aim being to integrate knowledge on ergonomics into shaping sitting solutions for active, working people. The whole production facility is imprinted by the challenge of good design. The chair producer's vision statement 'different and better' underscores this point.

The relationship started in the early 1990s, when SINTEF Production Engineering successfully implemented production control principles based on Kanban and JIT. In the first meeting with the company it became clear that the technological solution functioned quite satisfactorily, but some signs of a mismatch between technology and organization were visible.

We started our engagement in the company by conducting a conventional assessment of the enterprise. This was done through intensive interviewing of key management, trade union officials and shop floor workers. The social scientists played a dominating role in this work and the engineers functioned as key informants and discussion partners. This 'sounding board' cooperation worked well, both as a corrective to get as good as possible an understanding of the company and as a ground for mutual learning. Putting this information together in a report raised several important questions:

- How to handle great variations in the incoming orders, to avoid the production department running out of parts and to deliver orders at the appointed time?
- How to train the middle managers in organizational issues, since their background lies in ordinary production work?
- How to integrate the information system and the manual Kanban control system used in production?
- How to develop a smart way of organizing the production department according to some forms of semi-autonomous groups?

In the time after this initial assessment we continued discussing how to move ahead. It was difficult to find a point of entry that had an immediate resonance both for the local company and for the research team. After spending some time discussing this issue, the next step became a work climate survey. This was an activity with clear mutual interest, and it was based on work already done by the IFIM researchers. The questionnaire was adjusted to the local conditions and at this time the engineers in the group actively participated in the work. The survey focused on issues like job satisfaction, cooperation between management and workers, and on leadership.

The survey was quite successful. The highly motivating activity of encouraging the employees to respond and the process of sending out and collecting the surveys meant that a stunning response rate of 99.5% was achieved. The report was written and the findings were reported back. The feedback sessions were important, as they gave room both for general presentations to all employees as well as discussions with individuals. This interpretation of and deliberation on the findings was important both for the research team and for the company. We have now been asked to prepare a comparable survey for the total national and international organization.

In early 1997 the module's researchers held a seminar in the office chair producer's home town. The design of this seminar was intended to support the internal developmental process in the group by shaping a common intellectual ground based on concrete company experiences. We organized this workshop to include a company visit and spent ample time in debriefing our experiences. We also invited both the management and the union officials to participate in this event, thus really shaping the triangular communication involving company employees, engineers and social scientists. This discussion turned out to be reasonably successful, and a lot of energy was devoted to conceptualizing the next generation production system. We now seem to have started to pave the road that the three parties will build as the developmental process moves along in the company. Integrated in this process is the OD process within the research group itself.

The offshore construction yard

The offshore construction yard produces steel structures for offshore installations in the North Sea. The company is located in Central Norway and employs more than 1000 people. The competition on the offshore market is fierce, and is characterized by a continuous chase for new contracts. When we started our engagement the company had no new orders and the current production was coming to an end, the management was at that point talking of layoffs. Today, the yard has orders until the end of 1999, which is more than they have ever had before. The company is a production facility of a larger corporation, which in addition owns several offshore yards, engineering companies and other offshore related enterprises.

The first contact with the offshore construction yard was made in the summer of 1996, through a research project on enterprise modeling financed by the Research Council. This project, which was a joint effort between the company and SINTEF Production Engineering, had been granted money on the condition that social science researchers were included in the project. The project leader then got in touch with the Trondheim module, and we were included as sub-contractors. After some time, the research team accepted that the challenges within organization and management were equally important as challenges of enterprise modeling. Thus, organization development was established as a separate sub-project in the organization.

Since 1995 the offshore construction yard has been involved in a series of development efforts, aiming for process orienting the total workflow. The enterprise modeling project was initiated as a part of these efforts, together with organization development. Both projects were difficult to get off the ground, partly because of the lack of technology and manpower. Not very much had been done by the end of 1996. However, one engineer from SINTEF and a few project champions from the yard were impatient. They decided to get a few projects up and running in order to provide quick results. They initiated a pilot project in the iron cutting process, where a simple use of information technology proved sufficient to simplify the process of creating drawings, planning the cutting, and doing the actual manufacturing work. In arguing for this project, the project leaders dug out examples which proved how unnecessary complications created waiting time and difficulties in all parts of the organization. Simplification came to be one of the slogans of the project. This work was mostly done by engineering 'expert' involvement

The challenge for the organization is considered to be one of process orientation. This implies developing an internal and external organization to support the notion that all work processes can be carried out with a minimum of grievances. Main elements in the solution seem to be:

- To let engineering, supplies and the production of documentation and drawings be controlled by the needs of manufacturing.
- To manufacture sub-parts in a more industrialized way (from order production towards flexible mass production).
- Flexibility and reduction of uncertainty.
- The implementation of standards, especially in engineering and construction.
- A better flow of work, with fewer relays.

Enterprise modeling and organization development are important, necessary tools to achieve this.

The project aims at improving the flow from construction to manufacturing, to use computers, organize teams both in manufacturing and construction, create new roles for middle management, and increase participation in the development work. Finally, the aim is to diffuse promising results to other processes in the company.

The research team started its engagement by interviewing all participants in the process. These results were fed back to the interviewed people at the yard, the research team, the Iron and Metal Workers' Union, and the managers. In this respect our involvement is parallel to that with the office chair producer. This initial scan give us the basis for making sense of the company. On the other hand, there is another important difference regarding our engagement in the yard. When we were called in to manage organizational and leadership issues, we were expected to integrate these insights into the original engineering project. One way to support this process was the construction of a steering committee involving both engineers and social scientists. In the steering committee the module leader met the engineers and the company representatives, and through these discussions an agreement on integrating process orientation with organization was easily agreed upon. The real challenge was how to reach this goal.

The research team decided to work at the yard for two days a week. There they share the same office space as company people that were engaged in the process orientation project. The opportunities for cross communication were excellent, and mutual learning gradually emerged. In the first period the work was characterized mostly as parallel activity. The members understood what they were individually struggling to achieve, but few bridges were built that were directly linked to the project activity. However, there has been a certain strain between the action-oriented, quick results approaches of the engineers, and the more reflective and long-term approaches of the social scientists. This is visible among the researchers as well as among the organizational members. The company is in a hurry, because the next major construction project will be initiated in a few months, and it is important to get new routines and organization implemented in that project. Social science and enterprise modeling does, however, take more time than coming up with concepts for plant restructuring. In order for this to be investigated, there is a need to document the scientific process. This strain can be interpreted not only as differences in opinion, but differences in perceptions about humans and technology, rooted in the two paradigms of science and research. The challenge is to develop the research team and at the same time contribute to the development of the organization through shaping a fruitful dialogue through which all involved can learn.

The joint location of the research team is an important condition for achieving transdisciplinarity, but it is not enough. One successful arena that was created within the team was a de-briefing session held at the university. This meeting was called by the module leader in order to discuss the issues of integrating engineering and social science. The meeting was very constructive, as it gave room for a research-based reflection, aimed at figuring out how to make sense of the transdiciplinary view of the total project. This discussion led to a continued effort to integrate the different professional approaches, and to adopting a new strategy for creating a really integrated approach to process orientation. An interesting outcome of the same meeting was also a clear understanding of the importance of the researchers' integrity. To be competent players in the triadic relationship, it is also important to develop the genuine disciplinary and transdisciplinary knowledge base for the research team.

The aluminum smelter

The aluminum smelter is part of a larger Norwegian corporation. It employs some 900 people and is by far the dominant company in the local economy. It is located in a fjord on the western part of Norway. For the last 10 years the company has been successfully running a consecutive set of programs aimed at enhancing worker participation in the operation of the plant. The hierarchical structure has been flattened and health and safety issues have been a core focus in advancing the quality of work.

Our work at the aluminum smelter has focused on leadership training. We have been engaged in an effort in which organization development was integrated in the leadership development program. The core idea was to make the participants in the leadership development program take on the responsibility of running an OD effort in their own department. The experiences gained from these OD efforts would in turn be utilized as a basis for reflection and development in the leadership program. This program is still running, and we will continue working on it throughout 1997. The next issue we are to engage in at the smelter is to start working on centralization and decentralization problems linked to the corporation's effort to create a virtual company based on the five production units.

The maintenance staff of the Aluminum Smelter constitute about 10% of the total payroll. The research team from the Trondheim module includes a researcher from SINTEF Safety and Reliability. A sufficient and correctly designed maintenance standard leads to an elimination of technical failures and problems. This in turn makes maintenance invisible. The central ideas of our development work within this field are to make maintenance staff and the other part of the organization aware of the importance of maintenance, and make the performance of maintenance measurable. To make maintenance visible, it is necessary to develop a system of indicators that measure the maintenance performance.

Our continued engagement with the company will rest on issues of leadership development combined with organizational development and, hopefully, with a mutual engagement in the field maintenance.

The gas turbine component producer

The gas turbine component producer is located in a town some a two hours drive from Oslo. The company produces components for aircraft gas turbines. The company is tightly integrated into the production chain of its two large customers: Pratt & Whittney in the USA and Snecma in France. The two customers together own one third of the shares of the gas turbine component producer. The competition is generally fierce, but getting a contract implies a perspective of 50–70 years, including potential delivery of spare parts. The company is thus operating with a long time horizon, which again indicates long investment cycles. The manufacturing technology is quite sophisticated. The company's core competence is skills and experience in machining very complex materials.

The first introductory discussions with the management and union representatives took place late in the winter of 1996. This initial contact was followed by a company assessment, much like what we did in all the companies with which we were involved. The information was collected through semi-structured interviews with management and employees. In addition we used different documents such as annual reports and public relations material. These activities and discussions involved only company people and social scientists. The report focused on the following issues:

- The need to restructure the production.
- The development of labor–management relations to accomplish more cooperation, a better climate for cooperation.
- The need to start development/training programs for management at all levels (top, middle, etc.).

Obviously, this approach, only involving social scientists, needed to be compensated for. Our strategy for handling this was involving the whole research team in discussing the analysis and conclusions from the first assessment of the gas turbine company. This is a method which we frequently use within the research team. After the initial phase, when we struggled to find a shared language to create a mutual understanding of the field, this has turned out to be very interesting and fruitful. We have for a long time been in discussion with the company regarding how to proceed, and no final decision has yet been made.

The sausage factory

The sausage factory is a small Norwegian company with 60 employees. It has expanded from a staff of 20 over some years. The main challenge for the company is to industrialize the production, and thus to move from a small-scale, craft-based production. The management sees the implementation of a strategy for information technology (IT) as a key factor in the industrialization of the company. This includes the development of middle management, including the transition from an entrepreneurial management style to an organizational structure divided according to tasks and responsibilities. It also includes efforts by management to unionize the company. The understanding is that through development of a local trade union they will get a more stable and predictable work force. Today, approximately 50% of the employees are organized.

The contact with the Trondheim module was established through an earlier evaluation project. This led to the first discussions on how the module could help the company in the transition process towards modern industrial production. Through these discussions, the management expressed an interest in modernizing their information technology. One of the team members took on this task He worked with the company on developing its IT strategy. In this project, special attention was given to the implementation of the technology in the organization.

The company's approach was to try to accomplish successful implementation by re-organizing their already existing improvement groups. These groups were responsible for information technology investments, the use of IT, and competence development among the users. The group was also involved in handling the general information flow across departmental boundaries. We assessed the local information technology, and some improvements were suggested. Major issues were how to store and retrieve information in their network environments, their implementation of data acquisition in the production department, and their monthly reporting control parameters for the leader team. A general information strategy was designed based on this assessment.

The sausage factory is the only case in which the relationship between the social scientists and the engineers has not become an issue in the project. This is not because the transdisciplinary focus has not yet been established. The entry point for further development will probably be connected to technological change.

Summary of the research portfolio

The following table summarizes the main elements of our research agenda in the different participating companies.

As we can see from this table, the recruitment of companies was basically linked

Table 1: Overview over the companies

Company	**Recruitment**	**Problem Definition**	**Problem Statement**
Offshore Construction Yard	Research Council	Engineers, social scientists and company people in a joint process	1. Reorganization of work under business process redesign structure. 2. Enterprise modeling and organizational development
Office Chair Producer	Engineering project shaped new production technology, then a need for organizational change	1. Engineers & Company 2. Social Science-company 3. Engineering, social science and company	1. JIT and new organization 2. Semi-autonomous work groups 3. Redefining middle management
The Aluminum Smelter	Social science Ph.D. program	1. Social Science-company 2. Social science, engineers and company	1. Leadership development and organizational redesign 2. Integrated maintenance
The Gas Turbine Part Producer	Company contacted the module Research Council attempted link to a technological program	1. Social science-company 2. Research council and company	1. Process oriented group based organization 2. Leadership development
The Sausage Factory	Social science through an evaluation program	1. Social science-company 2. Engineering-company	1. IT strategy 2. Continuous improvement 3. Development of a local trade union

to previous engagement, either from the engineering side or from the social science side. One company was recruited through demands from the Research Council related to granting money for a technological project. One other company contacted the module to inquire about participation. In the problem definition phase it seems as if they started within one specific discipline, and moved from there through a new, opposing dyad to a triad involving all relevant parties. One exception is the offshore construction yard, where, even at the beginning, the project involved all members of the triad. Looking at the problem statements, there seems to be a major focus on leadership development, especially for middle management. On the other side, the dominating issues of research are integrated technological and organizational developmental strategies.

Company research teams

In the presentation of the companies we have shed some light on how the research teams were formed and worked in each specific company. Some of the researchers are involved in two such groups (and enterprises) and some are only involved in one. The work in the teams is partly based on a close dialogue with the enterprises and partly on the cooperation between the researchers involved in the attempts to produce new knowledge, which may cross the traditional barriers between the disciplines represented.

Except for one company, the team originated either in engineering or in social science, but in all cases it developed into a triad, integrating engineers, social scientists and company people in the same research process. The cooperation at company level has so far not created any great degree of tension. A very important design criterion was to bring the different disciplines together in a context of concrete problem solving. The expectation was that the reference to a common, shared reality would provide a fertile ground for mutual understanding. This seems to a very high degree to be the case. Almost without exception, the team members experience an interesting and stimulating learning process, giving insight into different ways of interpreting and framing problems that have been identified in common. This can, of course, be seen as the first step in creating a shared understanding of business development. The most systematic review of this kind of experiences was summarized after most of the Trondheim module members had participated in company assessments in a different large-scale project.[1] This work clearly benefited from learning through company work in the module, and might possibly have been much more difficult had it not been for the knowledge produced in the module. All in all, it seems likely that the integrated team approach has been very successful in creating the first shared visions on company development.

Plenary module cooperation

The module has 12 researchers enrolled. They represent a diversity of professional interests, knowledge, experiences and traditions. The main task has been to develop an organization which balances the needs of collective and individual development and learning inside the research group, but also the importance of balancing the inner dynamic processes in the research group with the dialogues with and work in the enterprises.

The module meetings

The module meetings are short meetings (e.g. 2–3 hours) or longer seminars (2–3 days) at which all the researchers within the module meet to exchange information, to present the development work within the participating companies, and to discuss theoretical and methodological questions of interest for the ED 2000 work. These meetings or seminars sometimes include (external) persons who are connected to the module on a more or less permanent basis. We have initiated permanent cooperation with outside professionals. Professor Davydd J. Greenwood of Cornell University has been actively engaged for a couple of years. We have organized semi-annual reviews and reflection processes involving Greenwood.

In the initial phase the plenary meetings functioned unsatisfactorily. The discussions were too general and no relation to a shared experience existed. This created internal stress in the group, and we redesigned the work to be more team-oriented, organized around each company. This turned out to be a more fruitful approach. A new effort to revitalize the plenary discussions was to combine them with visits to the participating companies. This is a quite promising strategy. We held a meeting this spring at the office chair producer's premises. At the end of the visit we had a joint discussion between management, the local trade union and the research team. This was a very fruitful discussion, dealing with the next generation production system. The cross fertilization between company people, engineers and social scientists created potentially interesting new views on how to organize the production system. This might seem to be a promising road to follow.

The network meetings

A strategy for extending the learning and communication in the module is the organization of network meetings. All participating companies and the entire research staff are invited to these meetings. The topics for discussion are based on the problem focus in each company. The fist meeting was very successful, and the companies showed great interest in continuing this work. Unfortunately, the company that volunteered to arrange the next meeting did not fulfill its promise. Unexpected circumstances meant that the meeting was not held, but we are now well under way to planning the second meeting. The network meeting opened up the possibility for all three parties to engage in a discourse on developmental issues. It also created cross contact between companies, and there have been visits between the companies as a result of this initial contact.

What have we learned

The process is still in its early stages, both when it comes to development activities in the companies and the internal learning process in the Trondheim module. Still, when reflecting upon our experiences, we have no difficulty in coming up with several examples of what we as researchers have learned that will improve our engineering and social science practice (the following concrete points are from a reflection session with some of the researchers):

- KANBAN production control systems: the social and practical limitations and possibilities of implementing this kind of system.
- The limits of hierarchical modes of organizing: the traditional kinds of authority, the function specific budgeting, the institutional frameworks are limiting in the same way as technology is.
- There is a mismatch between competence development and the highly structured, low-demand technology of mass production.
- In research methodology, there needs to be a relationship between getting a grip on the history of the organization and improving the current situation: one should not do development work without looking back into the past, and one should not map the past without having a developmental purpose.
- The same kinds of solutions are considered satisfactory within different scientific disciplines and for different reasons: teams in quality systems, zone-oriented manufacturing in production control systems, objects in information systems, groups in social systems.
- The conditions needed if technological solutions are to work are often defined in organizational terms: this is for instance the case for maintenance.
- We may change the technical solutions based on what we have learned from the organization.

A number of these lessons have of course been discussed earlier in the literature — within the socio-technical school, for instance. For us, however, it is important that the learning is experiential and experimental, and not theoretical. It provides a totally different basis for assisting development work in the companies, as we are better able to use the expertise of each of the parties: engineers, social scientists, and local actors.

Conclusion

The Trondheim module approaches enterprise development as an integrated effort combining professional skills in engineering and in social science. The cornerstone of our approach to enterprise development is based on an understanding that a developmental effort in organizations is not solely either an engineering or an organizational issue. Organizational changes can hardly be developed unless technological, structural and economic factors are considered, and *vice versa*. In a context of enterprise development our main position is that any change effort will inevitably effect other central issues in the company. Taking this point of departure, our action focus is integrated technological and organizational developmental activity. Our research focus is devoted to understanding the premises for multi-professional cooperation and on creating a shared frame of reference between engineers and social scientists.

A key point in our effort has been to construct arenas where engineers and social scientists together join in as a team to work in each company. A core element in this work has been to create a mutual understanding. This shared field of understanding is shaped by working in the same context and addressing the same problems, given the specific company reality. This mutual learning process is made possible through the involvement of management and unions. This learning triangle is a key factor, as the problem owners in the companies are the judges of the usefulness of the work, so to speak. Any principal discussion between different professional actors has to take account of how arguments are interpreted by the problem owners. This grounds the knowledge production in real life challenges, avoiding debates on matters of principal.

Note

1. This pilot project, Productivity 2005, focused on next generation production systems. The central idea was to create a holistic understanding of production systems. An important part of the project was the conduct of company assessments and the member from the Trondheim module played an important part in this.

References

Berger, P.L. and Luckmann, T. 1966. *The Social Construction of Reality*, Harmondsworth: Penguin Books.

Elden, M. and Levin, M. 1991. "Co-generative learning: Bringing participation into action research". In W.F. Whyte (ed), *Participatory Action Research*. Newbury Park: Sage (127-142).

Greenwood, D.J. and Levin, M. Forthcoming. *Democratizing Research: An Introduction to Action Research.*

Guba, E. and Lincoln, Y.S. 1985. *Naturalistic Inquiry* Newbury Park: Sage.

Gustavsen, B. 1992. *Dialogue and Development.* Assen: Van Gorcum.

Isaacs, W.N. 1993. "Dialogue, Collective thinking, and organizational learning". *Organizational Dynamics*: 24–39.

Johannesen, K. 1994. *Wittgensteins senfilosofi* (The Late Philosophy of Wittgenstein). University of Bergen, Department of Philosophy, Paper No. 42.

Kuhn, T.S. 1962. *The Structure of Scientific Revolutions.* Chicago: University of Chicago Press.

Latour, B. 1987. *Science in Action.* Milton Keynes: Open University Press.

Levin, M. 1995. "The quest for quality in participatory inquiry: A critical reflection on validity". Paper presented at Forum One, Hull University.

Levin, M. 1997. "Technology transfer is organizational development: An investigation in the relationship between technology and organizational change". *International Journal of Technology Management* 14(2–4): 297–308.

Monk, R. 1990. *Ludwig Wittgenstein The Duty of a Genius.* Harmondsworth: Penguin.

Wittgenstein, L. 1953. *Philosopische Untersuchungen.* Oxford: Basil Blackwell.

Chapter 7

Networking to Learn in Nordvest Forum:

Optimizing the Learning Cycle of a Regional Learning Network Organization through Action Research

Jon Hanssen-Bauer

Introduction

This paper explores the theoretical foundations underpinning the Nordvest Forum Module of the Enterprise Development 2000 Program. This module is organized by Nordvest Forum Inc., which is a regional learning network organization in Norway. First, the paper attempts to construct a theoretical model for inter-organizational learning, where the overall network objective is to strengthen the development processes in each of the participating firms. The second aim of the paper is to describe an action research approach that is congruent with the research questions and development purposes of the Module.

Nordvest Forum is a multi-firm regional learning network centered around the city of Ålesund in the northern part of the West Coast in Norway. The network was established in 1989, and 45 shareholding firms have set up Nordvest Forum Inc. to service the network and to manage its activities. Nordvest Forum's stated purpose is to improve regional competitiveness by upgrading the management capacity of the shareholding companies and other firms in the region.

In 1995 the board of Nordvest Forum Inc. decided to apply for funding by the Norwegian Research Foundation's research and development program, Enterprise Development 2000. The board's interest in the program stemmed from their recognition of a general need for more involvement and support from the network to its participant companies engaged in enterprise development. They saw in the program a possibility to build up relations with relevant research institutions and to set the challenges that were being experienced within their own region on the research agenda of these institutions. In short, the program could help develop the network by upgrading its competence in and support for enterprise development issues.

I have previously described Nordvest Forum as a regional learning network organization that was developed in response to increasing exposure by regional firms to competitive environments on a global market (Hanssen-Bauer and Snow, 1996). The northwestern coast of Norway has for centuries been an externally oriented part of Norway where the area's links to the sea have made access to the international markets traditionally easy. Fish processors, ship builders, and furniture manufacturers in the region are highly export oriented, and the region is recognized throughout Norway for its industrious, entrepreneurial, business-minded people. Companies are typically small by international standards, even when they are part of large Norwegian corporations, and the family-owned firm is a prominent feature of the area.

A successful response to increasing competition on the world market requires great organizational flexibility and adaptability, which are generated by the ability to develop and apply new knowledge quickly and consistently. This fact has urged many scientists to focus on how to create such ability in competitive firms. Their focus has often been on large companies and on how earlier 'giants learn to dance' (Kanter, 1989), or on how an organization's wide repertory of knowledge can be integrated in a flexible and effective way (Grant, 1996). The resulting literature on organizational learning is voluminous, presenting to the reader several models of learning processes and of how to organize and promote such internal processes.

The Nordvest Forum case presents a different set of related research problems than that of learning processes within firms. As the market challenges and changes in the business environment for the small companies call for improvements in the organizational performance, these small companies are exposed to many of the same challenges as bigger enterprises. But they find themselves in a situation where it is difficult to identify themselves with the descriptions and solutions suggested in the current literature. Suggested concepts for organizational development are designed for large organizations. The experiences of development processes are typically reported only from world class, global corporations. Both experience and concepts, therefore, need translation and interpretation in order to become relevant for the small firms on the northwestern coast of Norway. In addition, these small companies often lack fundamental resources, both internal and regional, to organize their own learning.

The owners of Nordvest Forum created the network as a response to this problem. The network would compensate for the small size and the ensuing scarcity of internal resources of each individual company. By banding together, they also felt they would become a more interesting partner for research and development institutions. The network could compete more successfully with larger corporations for these centers' resources and know-how in organizational development than each individual company was previously able to.

Hence, Nordvest Forum contributes to solving a problem of critical mass and resources by creating learning processes across firms in the framework of a network. Its challenge, then, consists of finding the best way to mobilize the network to promote such inter-organizational learning within the whole network *and* thereby strengthen the organizational change processes in the individual firms. On these topics the literature is less abundant. The Nordvest Forum development module is an experiment in optimizing the learning cycle of a regional learning network with this dual perspective, through building relations to social science research institutions.

In the following I present the various theoretical elements that are needed to construct a model of the learning cycle of the Nordvest Forum network. I then describe the current network activities and how they contributes to the cycle. Thereafter, the bases for the design of the enterprise development module and how it fits into the overall learning cycle are further explained. The last section of the paper describes the action research model that is applied in the module.

The Nordvest Forum learning cycle

An attempt to construct a model of the Nordvest Forum learning cycle has already been published (Hanssen-Bauer and Snow, 1996). This model draws theoretically on various schools within the organizational learning literature. As these schools focus on various dimensions of organizational learning, it is necessary for our purposes to combine elements in order to obtain a more holistic model and to expand some elements so that the resulting model can guide our observations of the learning processes within *the network*.

The starting point is the belief that knowledge is both tacit and explicit or codified (Nonaka, 1991, 1994), and that it consists of the individual know-how, skills, and beliefs that enable members of an organization to make distinctions and ascribe meaning to observations (von Krogh, Roos and Slocum, 1994). As knowledge is contained in individuals, such individually held knowledge must be mobilized into focused, collective action to be useful in organizations (Grant, 1996). An organization can be conceptualized as structures and means to enforce coordination and focus on the individuals. The elements that structure the organizational behavior into patterns consist of values, norms, routines, rules and operational procedures and systems. These elements have been created or invented by the members of the organization, based on their combined and shared experience or knowledge, and are gradually institutionalized over time to become the organization's culture. The culture offers to the actors a set of models of the outside world, consisting of theories that confer order and meaning on it in a way that guides their actions, even if the theories are not always explicit. These models are normally internalized by the actors, at least to a certain degree, and are taken for granted as right, good or aesthetic, as they have proved their efficiency in practice, i.e. in guiding the organization's behavior and in solving its problems internally and externally.

In changing environments — i.e. those that regularly require the creation and application of new knowledge — the firms that have the ability to institutionalize various types of specialized new knowledge quickly and efficiently into their operational routines achieve superior returns. I use the term 'organizational learning' to refer to the process through which firms transform new knowledge into changed, and hopefully improved, organizational practice. The concept, therefore, includes both individual learning and the institutionalization of the new knowledge into the shared knowledge structure of the collectivity.

How, then, is new knowledge generated in firms, and how is it transformed into improved organizational performance? According to the information-processing view, a company's stock of knowledge is created and expanded when information is obtained, processed, and stored for future use. The learning process may be described in terms of its stages and its effectiveness in each stage (Daft and Huber, 1987; Huber 1991). In my opinion, this view focuses essentially on *passive* acquisition and processing of information which centers on what is given to the organization. It tends to reify knowledge, as it gives inadequate consideration to how knowledge is *created*. A more proactive view is presented by scholars like Nonaka, for whom learning is an innovative process in which the organization creates and defines problems and then actively develops new knowledge to solve them (Nonaka, 1994). Combining these two approaches, we can say that learning occurs in stages within a context of purposeful interaction by individuals seeking to understand, and trying to adjust to, fluctuations in their environment, or attempting to recover from misguided previous perceptions or breakdowns of interpretative models of the world.

Hence, learning is an activity that can be depicted as experimentation in the real world, as the individuals generate new knowledge by doing. To influence and change organizational performance, however, the individual experience must be shared among the various members of the organization and the new knowledge must be accepted as justified, before it is eventually internalized into new norms and changed behavior.

To learn, individuals must explore and share their knowledge and interpretations and through this process develop new and common conceptualizations. As part of the knowledge is tacit, and even taken for granted and not open to questioning, the learning process implies codification and even confrontation of views, before they can be combined into new models. These models, as they develop, tend to hide their origin once again, and become internalized, partly as tacit knowledge, partly as orientations that are taken for granted, and, only partly, as explicit or codified assumptions (Nonaka, 1994: 19). The learning process, therefore, implies not only passive sharing but active apprehension and legitimization, which implies a restructuring of the knowledge of both the individuals and the collectivity concerned. As a new element or concept is being apprehended by the collectivity and is inserted into previous knowledge and models, it gradually builds up density. It becomes more and more meaningful for the actors, and will eventually be internalized as legitimate and firm or solid knowledge.

Local knowledge and assumptions are externalized and confronted with practice most effectively in change projects. In such projects, new concepts are formed on the basis of shared experience. I will use the term *focused experimentation* to distinguish such processes both from random and unintentional learning and from individual experiments. Through focused experimentation, old models and assumptions are opened up and deconstructed, and new concepts and theories are searched for. The process through which these new concepts and meanings build up density as they are gradually incorporated into the local repertoire of legitimate knowledge will be termed *cultural crystallization* below. Through the cultural crystallization, the local knowledge is reconstructed in a way that also restructures corporate behavior.

The literature offers several models of how sharing of perceptions and development of new interpretative models can occur, and of the various factors that inhibit or favor learning. Paradoxically, organizational learning must build on commonality and at the same time be nurtured by diversity. The essential characteristics of a favorable learning environment include committed interest, trust, and shared language and cognitive maps for interpreting information (Stata, 1989). Organizational learning is also facilitated if knowledge that is taken for granted is challenged by diverse and novel information provided by actors having different experience and views. Broad and open networks help to widen the mass of information accessible to the actors (Granovetter, 1982). Learning can best happen in arenas that are structured so that an organization's members adjust their diverse views and reach consensus on the value and legitimacy of new meanings and approaches in *dialogues* (Gustavsen, 1992). It is also possible to argue for the value of *paradigmatic examples*. Such examples are stories about other companies that function like metonyms or icons for whole packages of problems and solutions, and that advocate their own inherent legitimacy as proven truth. In Norway we refer to ABB, to Toyota, to Microsoft, to Volvo, etc., and citing the example functions more like an argument than as illustration of theory.[1]

The process of constructing collective meaning is enhanced when an organization's members have the opportunity to engage in self-managed problem-solving with adequate access to others' experience and backgrounds. Often, learning requires the establishment of flexible organization structures that enable members to cut across previous communications channels, and the implementation of nurturing managerial styles (Nonaka, 1994). Special conferencing techniques have been developed to foster favorable environments for dialogues and collective searches (Emery and Emery, 1978; Elden and Levin, 1991). The term *development organization* has been used for structures combining arenas, conferences and dialogues to promote organizational change in an organization in a systematic way (Gustavsen, 1991).

A learning network can be analyzed as a development organization that encompasses a number of firms and that links individuals to a learning agenda without participating in the same business. An ideal model of a network learning cycle is shown in Figure 1. The network learning cycle may be understood as the transformation of information acquired by individuals, through interpretation and consolidation, and, eventually, to crystallization into a restructured local knowledge systems of the participants. In the model the cycle has been divided

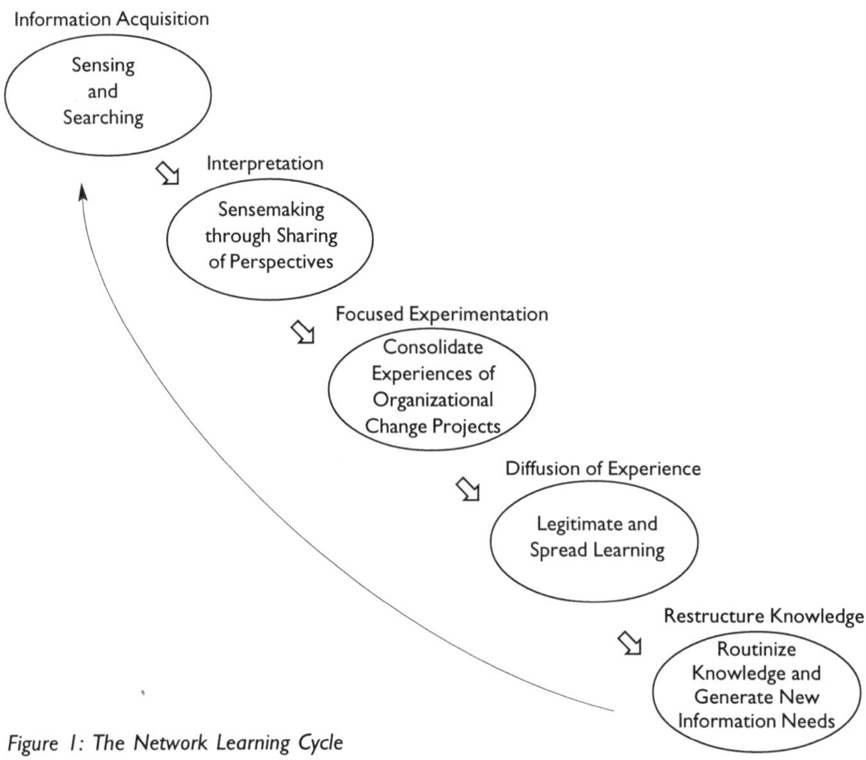

Figure 1: The Network Learning Cycle

into five stages in order to analyze how various activities and arenas in the network contribute to learning on different levels in the network. To some extent the stages can better be described as functions or processes that must take place in order for learning to occur. This idealised model should not be taken to imply that the cycle is neatly staged as a linear process. There are many feedback loops and learning may involve many of the described processes running in parallel. A network learning cycle may be distinguished form both personal and intra-organizational learning in terms of differences in arenas and in how the actors are mobilized and participate.

Nordvest Forum and its activities

In this section I briefly describe how the activities of Nordvest Forum contribute to the effectiveness of the learning cycle. The explicit purpose of the network is to contribute to enhancing the regional competitiveness. According to the owners of the network, this overall aim will be reached if the activities of the network are designed in such a way as to increase the management capability of the participating firms, to increase the company flexibility and effectiveness, and to contribute to increased inter-company cooperation.

There are five kinds of activities or products (see Figure 2). The main activity has so far been the Management of Change program. This is a management training program that has been particularly designed for Nordvest Forum to respond to the needs as defined by the owners of the network. It consists of five one-week modules covering a wide array of management skills and issues, running over 1½ years. Each class is attended by 20–25 managers, and the presenters and facilitators are recruited from various universities, from R&D institutions, and from advanced companies. To date, more than 200 managers from nearly 100 enterprises and public institutions in the region have passed the program.

Another activity that was established early on, was a forum for senior executives. This is a two-day gathering for senior executives in the region. The primary objective of this forum is to provide a meeting place where executives can hear presentations on international subjects and discuss issues of importance to the companies in the region. The meetings also provide a convenient setting for personal and professional networking. Today 19 such meetings have been held over seven years. Nordvest Forum also offers custom-made training programs for enterprises, and they organize special seminars and workshops on issues that have been defined as interesting by the participants of the network. Nordvest Forum also facilitates relationship building among the managers and the enterprises in the region.

Each of the owners has provided the same amount of capital to establish the network and holds one share in the company Nordvest Forum Inc. The owners

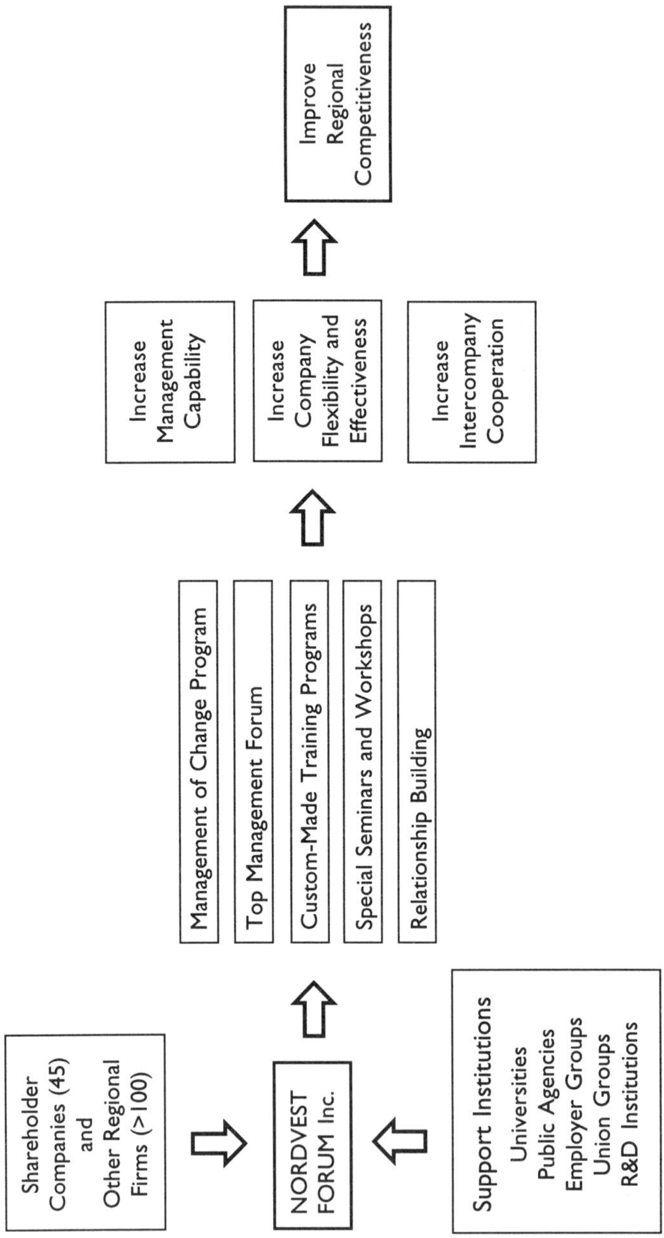

Figure 2: Norvest Forum: Components, Activities, Objectives, and Purpose

have formed an executive board that provides the strategic direction of the network. In addition, a group of professionals have been recruited to form what the network calls the Development Group, i.e. the body that actually helps design and renew the activities of the network. Only two persons are employed full time, since all the necessary professional input is provided by Nordvest Forum's network of professionals and institutions, both in Norway and internationally.

At this stage it may be useful to give a comment on the nature of the Nordvest Forum network's membership. On the one hand, I have so far described the organizational construct as a network *organization* when referring to the service company established by the company owners. It makes sense to describe this company as the network construct *per se*, and, eventually, treat the other companies around Nordvest Forum Inc. as its clients. On the other hand, the participants themselves look upon network building among the users of the network's services as an important aim in itself. Nordvest Forum's initial brokers explicitly wanted the participants in the Management of Change Program to build personal networks that could be mobilized to support themselves, as managers, and as resources for their companies. Such network building undoubtedly does occur, although the extent to which it occurs, and how much networking can be said to include the *companies* and serve business purposes may be a matter for discussion (see Snow and Hanssen-Bauer, 1993; Johannessen, 1996).

If we take the latter approach and consider the various forms of relations and exchanges created and continued among the various users of the network, we are struck by three important phenomena. First, 'membership' of the network has a very fluid status. You can 'walk in' and 'walk out' of the network according to your needs and interests. There are no membership fees, and most activities are open to everybody who 'feels at home' and wants to participate. Fees are paid on a cost basis for each individual activity. Second, the Nordvest Forum Inc. has a set of structured activities that are continuous, such as the Management of Change Program. These structured activities serve as a 'door' to the network, and people and companies can choose to involve themselves whenever they want, at a relatively low cost and with few binding engagements. Third, the users of the network vary in the degree of continued participation in network activities. There is a core of permanent participants who always show up, a large number of persons from the companies having 'been to Nordvest Forum', as they put it. Others are more peripheral or occasional participants. To some extent the core participants are to be found among the owners, but not exclusively so. It seems that senior executives make greater use of the network's potentialities than lower ranking employees in the regional firms (Johannessen, 1996). In sum, these characteristics means that the organizational construct fits various definitions of the 'network' concept, and it is theoretically interesting to explore the network's functioning and how it actually links the various regional actors into a common web.

Let us now look at how the activities and arenas are linked to the network learning cycle as defined above. The *information acquisition* is strengthened through the lectures in the management training program, through professional networking, and through the senior executive forum. In addition, the links to the R&D institutions are used to mobilize support and input as required. However, it was clear to the owners of the network that this would be a limited function if they did not design arenas for facilitating the *interpretation and making sense* of the information acquired, so that it could be translated and apprehended. Therefore, the management training program makes extensive use of discussions and case analysis in groups that are stable throughout the duration of the course. The special seminars also provide an appropriate arena for working through new information in depth.

Each participant in the management training program is expected to undertake a change project in their own company, under the supervision of a senior executive in the region. This activity falls within what we have termed *focused experimentation*, as it should in principle help consolidate the individual's learning and provide a mechanism for extending the learning to the participants' enterprises. The experience from these projects is discussed in the whole group in order to spread the learning to the other participants. Some of the projects are also later added to what could be called the network's development project library, as they are used as learning cases for later courses and workshops. The function of these activities is both to *diffuse the experience* and to *legitimate the new knowledge* by studying how it can actually be incorporated into local businesses as *reconstructed knowledge*.

This description could be extended, but it suffices here to claim that the total structure of the network activities can be analyzed as a regional development organization for the participating enterprises. After a few years of operation, an evaluation of the network activities showed clear indications of positive results on the management capacity of the participating firms (Snow and Hanssen-Bauer, 1993). This success has gradually been translated into a demand for renewal and for more learning.

Today, Nordvest Forum is acknowledged as having developed one of the best networks in the country for quickly accessing management-oriented professionals and institutions in Norway, Europe and the United States. It has established links with the central research institutions in Norway, such as the universities in Trondheim, Bergen and Oslo, to various research institutes in Norway and Scandinavia, to consultancy firms, as well as to departments responsible for organizational development in large Norwegian and Scandinavian corporations. The external network includes public institutions, as well as employer and union organizations in Norway. One important linkage is with the Institute for Management Development (IMD) in Switzerland, where Nordvest Forum Inc. is a Business Associate. That relationship gives the network companies access to the 'best practices' of notable global corporations, and some of the network companies have been used as cases for the development of new concepts and theories that have been widely disseminated (Vollmann, Cordon and Raabe, 1995, 1996).[2]

The module's main research problem

Analytically, we can differentiate between two levels of the learning cycle. First, we can analyze the learning as it occurs among the individual participants in the network activities. We can isolate aspects that can be argued to facilitate the transformation of personal learning into organizational learning within the participants' respective companies (Hanssen-Bauer and Snow, 1996). It is both interesting and challenging, however, to change vantage point from the network activities and the individuals to the participating enterprises. On this second level we focus on the multitude of learning cycles occurring between and within the network enterprises in the region, in order to study their inter-linkages and how they are intertwined and nurture each other, as well as how they form independent trajectories. This will enable us to describe and analyze more fully how the network activities translate into increased flexibility and effectiveness in the enterprises. We need careful empirical observation of organizational learning and development projects in selected enterprises, as well as documentation of the various linkages and channels that exist between these enterprises and the network activities. In this way we can isolate the factors that enable the integration of knowledge across the various organizations.

As more than 200 executives have participated in the Management of Change Program, the network has had an impact in creating and diffusing a stock of concepts and ideas, models and interpretations among the managers in the region. As core participant companies usually allow several of their senior managers to follow the program, the management teams of these companies have gradually built up a common understanding of the modern management and organization paradigms. Some of these companies have implemented internal add-on workshops that help take the organizational learning even further, changing practices in their firms. The organizers of the program believe that they are witnessing a shift in the regional management culture, since the reactions to modern organizational concepts are widely different now from the ones given by the first classes. The Top Management Forum has contributed to this shift, and new management concepts now seem to be taken up and circulated as rapidly in the region as in other business and management circles in the world.

Organizational development projects occupy a strategic position in the Nordvest Forum learning cycle, as they constitute a major mechanism that is believed to link the individual learning processes of the participants in the network activities to the learning processes of the companies. A key part of the Management of Change Program is a management development project, which is mandatory for each participant. Each project has a mentor, an experienced manager from the same company or another company in the region. The participants in the program are expected to apply the theoretical concepts and skills in the program in these projects in their own companies, and to function as each others' consultants as the progress and experience of the change projects are presented to their classmates. Further, the methods and results of the change projects are communicated to the other Nordvest Forum companies in various ways, particularly in subsequent sessions of the Management of Change Program where earlier participants are invited to present their projects as cases for learning. That approach has helped to build a stock of knowledge and experience which is shared by many firms in the region.

However, the focused experimentation seems to be the weakest point in the regional learning cycle. It has two weaknesses. First, the practice and depth of the current change projects in the region limit the learning potential. To some extent, more far-reaching project experience is 'imported' in the form of anecdotes or presentations in seminars. But, as is often the case, the value of such implants are often disputed by the local culture. There is therefore a need to strengthen the capacity of managing change projects in the participating firms. Second, the richness and rigor of the descriptions of the ongoing or past change experience in the region would gain from improvement. The interest of the Enterprise Development 2000 Program is that it enables us to address both these weaknesses at the same time. The Module will provide both for researcher participation in local development processes, and for scientific reporting from the same processes based on participant observation or action research.

In sum, the core research problem for the Nordvest Forum Enterprise Development 2000 Module is to develop a more comprehensive model of how a network learning cycle can be designed to encompass the linkages between the network and the individual companies. The module researchers will observe and participate in activities at the network organization level to support development processes in the participating companies. They will also follow and contribute to cases of enterprise development in member companies. This way, the Module learns how the network learning processes unfold in practice, and can analyze mechanisms that contribute to shaping the competitive business environment of this particular region. In other words, the Module requires an action research design that reflects the duality of the research problem.

The action research design

Action research can be distinguished from other types of research by the fact that researchers and practitioners engage in a joint process to investigate a given problem. The objective is both to identify workable solutions to the problem —

based on a merging of local knowledge and general assumptions and guidelines from theoretical knowledge — and to create new learning and knowledge during the process. At the core of the coalition formed between the researchers and the practitioners, therefore, is a particular social contract or relation that is both symmetrical and responsible (Van Beinum, 1990), as opposed to the expert intervention and to the non-intervening, observing researcher. The participants from both sides are involved in a development effort that can be described as an experiment in a local context, presenting many of the same learning characteristics as the focused experimentation described above.

The model of action research as it evolves over time is described in Figure 3. Action research is best described as a process that is cyclical in nature, but where the content and the problems researched changes over time as a consequence of the process. Therefore, action research takes the form of a '*hyper-cycle*',[3] to borrow a term from biology to describe a process that is cyclical, but not in the sense that anything is repeated a second time, since most components change as a result of the process itself. What characterizes scientific knowledge is the fact that it has been generalized through decontextualization or distancing the local experiences from its particular context. In this form, the generalizations, or

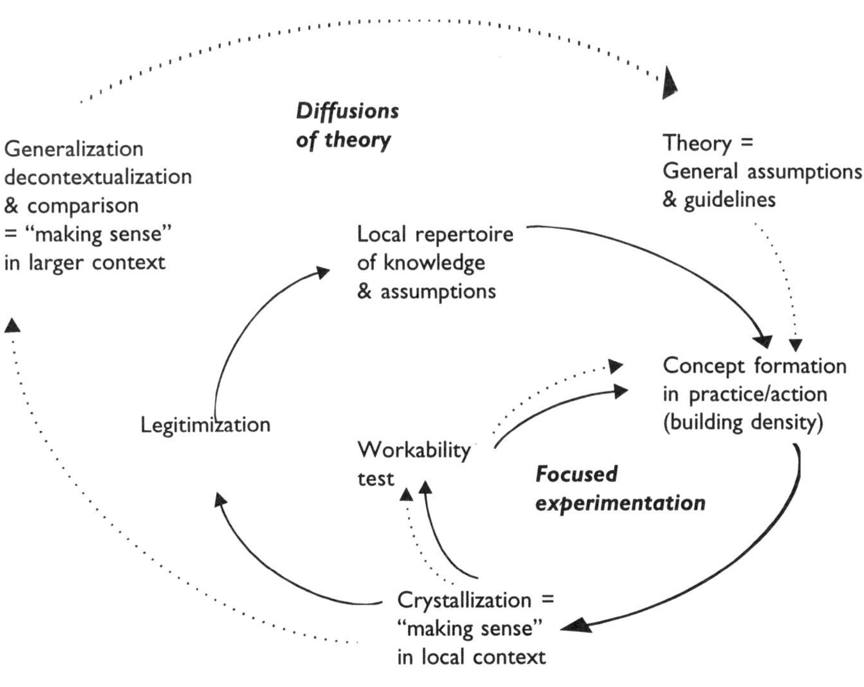

Figure 3: Action Research Hypercycles

theory, can be carried over to new contexts as general assumptions and guidelines for action. As practical models for action they are poor, however, unless they are brought to bear on a given context in juxtaposition with contextualized assumptions and guidelines drawn from the repertory of local knowledge. But the essence of research is precisely to test the general value and relevance of theoretical concepts through applying them in new contexts and observing their success in making sense of new experience.

On the other hand, what characterizes the local knowledge is that it derives meaning from the concrete situation, and, even if it functions as a model for action within this context, it must be stripped for the contextual 'thickness' (Geertz, 1973) before it acquires general significance as theory. General concepts and experience from other contexts are essential to create the necessary theoretical distance to allow this stripping to occur (Agar, 1986; Glaser and Strauss, 1967).

The strength of action research is to provide a paradigm for designing arenas for development coalitions (Engelstad and Gustavsen, 1993; Gustavsen, 1997) of researchers and practitioners to collaborate on focused experimentation based on both kinds of knowledge in the same learning process. The learning cycle consists of applying local and generalized knowledge to a development project in a way that favors the formation of concepts to account for the experience. Such concepts build up density and thickness until they eventually crystallize and are incorporated as useful elements to make sense of the experience within the local context. As they are shared and prove to work for more of the local actors, they become integral part of the local repertory of knowledge. It is the role of research, then, to generalize the learning through a description that allows for comparison with the experience from other contexts in a way that builds theory.

This model of the action research learning hypercycle has guided the design of the development and research arenas for and the organization of the Nordvest Forum module of the Enterprise Development 2000 program. The module consists of two levels of focused experimentation. First, the level of the regional learning network itself, where the module is embarking on several projects focusing on the efficiency of the learning network itself. Second, five clusters of companies engaged in development projects are participating in the module. Each of the clusters hosts a doctoral student in a discipline relevant to organizational change. The students serve as a 'significant other' to the companies in the local development processes, and from this position they carry theoretical elements from their studies to the companies and, in the other direction, they take situated problems for reflection back to the universities. The theoretical elements are not used as expert input, but as elements to stimulate local joint reflection and distantiation. The concrete problems are used in the universities as cases for reflection that serve to contextualize the theoretical concepts. As the students will follow the development processes within their clusters over three years, they will write their doctoral theses based on data from these cases.

A team of senior researchers is linked to the module on a part-time basis. Some of these are responsible for the doctoral programs at the universities where the students of the modules are enrolled. They will be the natural advisors to their students. In addition, both these teachers and the other senior researchers will conduct their own research projects, directed towards different aspects of the regional learning network and the development practice in the participating companies. The seniors will be mobilized as resources both for the network programs and for the company development projects. The company projects are highly varied. The companies are presented in Table 1.

The participating companies are drawn from different industries and they are confronting different challenges in their markets. They are also different in size and in organization, as well as corporate culture and management traditions. In a sense they represent the variety of the businesses in the region. But they are

Table 1: The Company Clusters in the Nordvest Forum Module

Cluster	Companies	Locality	Turnover (in mill. NOK, '93)	Employees (in 1993)	Industry
1.	Stokke Group	Ålesund	360	394	
	*Stokke Fabrikker AS	Skodje	189	155	Furniture
	*Vatne Møbler AS	Haram	28	51	
	*Westnofa Industr. AS	Rauma	127	132	
	*Møremøbler	Ørsta	90	125	
	Ekornes-konsernet	Sykkylven	533	711	
	*J.E. Ekornes As	Sykkylven	276	319	Furniture
	*Ekornes AS Stranda	Stranda	82	72	
	*Ekornes Fetsund AS	Fet	81	96	
	*Ekornes Grodås	Hornindal	19	54	
2.	Mørefly[1]	Giske	132	85	Air transport
	Ulstein Hotell	Ulstein	22	37	Hotel
	Møre Trafo	Sykkylven	136	128	Electro-mechanic ind.
3.	Domstein Group	Måløy[2]	1021	486	Fish processing
4.	Glamox Group	Molde[3]	737	953	Electro-mechanic ind.
	Moxy	Fræna	224	140	Prod of heavy dumpers
5.	Ulstein-gruppen	Ulstein[4]	2453	2593	Ship-building
Sum			**5851**	**5654**	

1. The company has been bought up and integrated into another air carrier during the project time.
2. Several companies in Sogn og Fjordane, and in Sweden.
3. Companies in Molde, Svelvik and abroad.
4. Several companies in Møre og Romsdal, the Bergen area, Kongsberg and abroad.

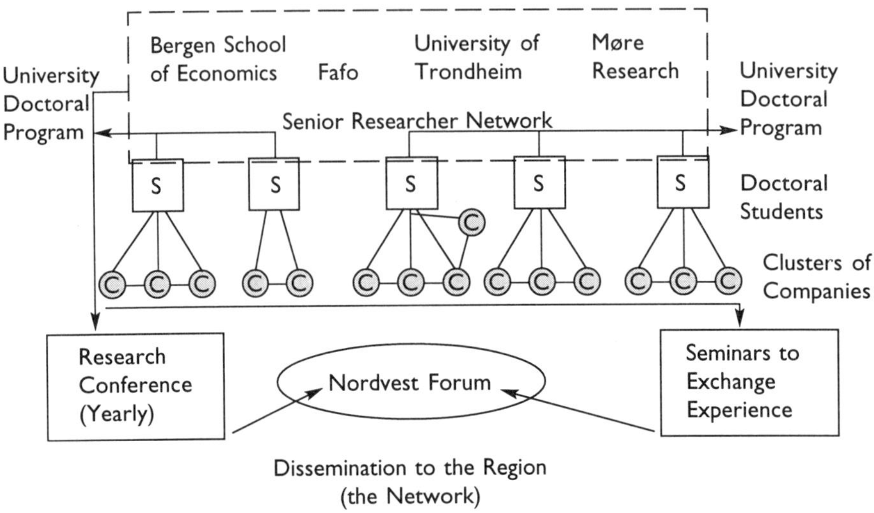

Figure 4: The Nordvest Forum Module

all among the core members of the Nordvest Forum network. As active members of the network, it should be possible to trace and locate the linkages and channels of information and learning being forged between the companies and the network. As active members, they will also themselves take part in the dissemination of their learning through the network. The doctoral students, as they are employed by Nordvest Forum Inc., will be involved in network activities as required, and in this way they will help to disseminate development experience.

In order to facilitate the academic part of the research process, the module has created an academic forum to which other researchers will be invited to take part in discussions of the results and data obtained from the Nordvest Forum. In order to facilitate the crystallization process and to support the development projects, the module will create a forum for sharing experience and raising problems of common concern among the participating companies. The organization of the module is shown in Figure 4.

Conclusion

In this paper I have described the theoretical considerations and previous research experience from Nordvest Forum underlying the establishment of this particular module of the Enterprise 2000 Program. I have described the Module program as an action research project aiming at intervening on two different levels of the regional learning network, i.e. on the level of the enterprises and on the network level. The design of the project incorporates the need to provide research-based experience on cases of business development, and at the same time to explore current practices, new experience and local knowledge that guides enterprise development, for further learning and dissemination.

The module has been designed for these combined purposes. As the module is managed as a network in itself, it is an experiment in action research that deserves some reflection as the project activities proceed. By way of an example, the module has built up doctoral projects in a new form of organizing that may influence the way the universities think about the relation between theoretical studies and practical involvement in business life. Hitherto, students have normally studied corporations; they seldom participate in internal processes and development work. We do hope that participation in internal processes and following these processes over an extended period will better prepare 'our' doctoral students for future careers than those who follow ordinary university lectures only. We also hope that their practical experience will color the way they read and, eventually, help to bring more of the daily business challenges into the classrooms. We already know that the new links established between the regional firms and the universities have created more interest for this part of Norwegian work life in the educational institutions. Therefore, some change will probably be detected by future evaluators of the program.

Future evaluators will also find that the number cases of enterprise development used for reflection and inspiration by regional managers has increased and that the cases are richer. These cases will exist in the form of academic texts, as illustrations in the courses and training programs, as anecdotes, and as paradigm examples in the local repertory of knowledge. Will the module help to change the region's business life? Will the companies be more effective and productive? Will they become more democratic and structured for better use of the total stock of knowledge among their employees? Will they be better able to manage change? To answer these questions will be very interesting, but they remain challenging questions for the future.

Acknowledgements

The author has worked with the Nordvest Forum network since 1989, while holding a position as project coordinator for the Norwegian Work Life Center (SBA). The network was supported by this national program for business development in Norway during the period 1989–1993. SBA was funded by the major work life organizations and the Norwegian government. A first evaluation of the network by the author and Professor Charles C. Snow (the Smeal College of Business Administration, Pennsylvania State University) was funded in 1993 by the SBA and the Norwegian Research Foundation. I am indebted to SBA and its director, Thoralf U. Qvale, as well as to Professor Snow, for many of the ideas integrated into this paper. The module idea and concepts were developed in close cooperation with the then director of NVF Inc., Knut Åsebø, and Professor Morten Levin at the University of Trondheim. Researcher Sveinung Skule, my colleague at Fafo, has participated in the project since its inception and has provided valuable comments on this paper. I am also indebted to research director Jon Pedersen, Fafo, who generously provided constructive criticism of the paper.

Notes

1. I am indebted to Sveinung Skule for making this point.
2. Haakon Raabe is one of the doctoral students in the Module. He developed his theoretical views while staying at IMD in Lausanne as part of his doctoral program. He has been working with the furniture industry (Cluster 1 in Table 1), in particular with Ekornes, where the concepts of Supply Chain Management have guided their development efforts. As a result, the Ekornes company has gained wide recognition as an innovative company.
3. The model was developed during the 'Scandinavian Based International Action Research Development Program (ACRES)', Sweden, 1992–3. I am indebted to my working group, with which I first discussed these ideas.

References

Agar, M.H. 1986. *Speaking of Ethnography*. Beverly Hills: Sage.

Daft, R.L. and Huber, G.P. 1987. "How organizations learn: A communication framework". *Research in the Sociology of Organizations* 5: 1–36.

Elden, M. and Levin, M. 1991. "Cogenerative learning: Bringing participants into action research". In W.F. Whyte (ed), *Participatory Action Research*. Newbury Park: Sage, 127–142.

Emery, M. and Emery, F.E. 1978. "Searching: For new directions, in new ways, for new times". In J. Sutherland (ed), *Management Handbook for Public Administrators*. New York: Van Nostrand Reinhold.

Engelstad, P.H. and Gustavsen, B. 1993. "Swedish network development for implementing national work reform strategy". *Human Relations* 46: 219–248.

Geertz, C. 1973. *The Interpretation of Cultures*. New York: Basic Books.

Glaser, B.G. and Strauss, A.L. 1967. *The Discovery of Grounded Theory: Strategies for Qualitative Research*. New York: Aldine.

Granovetter, M. 1982. "The strength of weak ties: A network theory revisited". In P.V. Marsden and Nan Lin (eds), *Social Structure and Network Analysis*. Beverly Hills: Sage, 105–130.

Grant, R.M. 1996. "Prospering in dynamically-competitive environments: Organizational capability as knowledge integration". *Organization Science* 7(4): 375–387.

Gustavsen, B. 1992. *Dialogue and Development. Theory of Communication, Action Research and the Restructuring of Working Life*. Assen, Stockholm: Van Gorcum and the The Swedish Center for Working Life.

Gustavsen, B. and Hofmaier, B. 1997. *Nätverk som utvecklingsstrategi*. [Network as Development Strategy]. Stockholm: SNS Förlag.

Hanssen-Bauer, J. and Snow, C.C. 1996. "Responding to hypercompetition: The structure and processes of a regional learning network organization". *Organization Science* 7(4): 413–427.

Huber, G. 1991. "Organizational learning: The contributing processes and the literature". *Organization Science* 2: 88–115.

Johannessen, H.A. 1996. *Utvikling av nettverk mellom ledere. Erfaringer fra Nordvest Forum* (Development of networks among managers. Experience from Nordvest Forum). Dissertation. Trondheim: University of Trondheim.

Kanter, R.M. 1989. *When Giants Learn to Dance*. New York: Simon and Shuster.

Nonaka, I. 1991. "The knowledge-creating company". *Harvard Business Review* 69: 96–104.

Nonaka, I. 1994. "A dynamic theory of organizational knowledge creation". *Organization Science* 5: 375–387.

Snow, C.C. and Hanssen-Bauer, J. 1993. "Nordvest Forum evaluation study report". Oslo: The Norwegian Work Life Center (SBA).

Stata, R. 1989. "Organizational learning — The key to management innovation". *Sloan Management Review* 30: 63–74.

Van Beinum, H. 1990. "Observations on the development of a new organizational paradigm". Paper, Seminar on Industrial Democracy in Western Europe, Cologne.

Vollmann, T.E., Cordon, C. and Raabe, H. 1995. "From supply management to demand chain management", *IMD Perspectives for Managers* No. 9. Lausanne.

Vollmann, T.E., Cordon, C. and Raabe, H. 1996. "Supply chain management: Making the virtual organization work". *IMD M2000 Executive Report* No. 19, Lausanne.

Von Krogh, G., Roos, J. and Slocum, K. 1994. "An essay on corporate epistemology". *Strategic Management Journal* 15: 53–72.

Chapter 8

Enterprise Development in Networks

Tor Claussen

Networks, quality management, business re-engineering and flexible specialization are some of the 'in' terms associated with new trends in the development of the business environment. In particular the utilization of network collaboration has been recognized as an important factor in achieving competitive advantage (Porter, 1990; Williamson, 1987; Granovetter, 1985). Furthermore, the development of networks is also an effective way of establishing learning organizations in view of its inter-organizational characteristics (Senge, 1991). In this paper we describe how enterprises collaborating in networks can acquire knowledge and competence through participation in joint enterprise development activities. Collaborating in network activities enhances the companies' awareness of new challenges and possibilities. The contribution that action research can make in building the necessary trust for enterprise development will be emphasized.

ED 2000 in the south west part of Norway: project focus

ED 2000 at Rogaland Research (Norwegian: RF)[1] focuses on the challenges that confront enterprises and networks when putting into practice international management concepts concerned with improving quality and organizational effectiveness, such as TQM and BPR[2], in enterprises characterized by a Scandinavian model of enterprise development.

We take the Scandinavian model to mean a business practice in which unions participate as an active part in enterprise development (Schiller *et al.*, 1993; Gudmundsson, 1993. See also: Quale and Øverland, 1996; Øverland, 1995). This Scandinavian type of collaboration is regulated by national legislation and local agreements (Venneslan and Ågotnes, 1993).

In ED 2000 at RF the labor market parties thus provide a very significant context for an enterprise development practice in which management concepts aimed at improving quality and organizational effectiveness are being introduced. One of the main challenges in this research project is to utilize the advantages from both new management and business concepts and the Scandinavian model of worker participation in such a way that it will facilitate *new forms of industrial democracy which will enhance the competitive advantage* of the participating companies.

The objective is to initiate intra- and inter-organizational development processes that will enable the enterprises concerned to operate effectively in an international, competitive environment. The project is aimed both at internal enterprise development and towards network relations.

Network collaborations[3]

There are three networks in ED 2000 at RF:

- The Technology for Sunnhordland (TfS) network
- The Technological Cooperation (TESA) network
- The SYNERGI network

The Technology for Sunnhordland (TfS) network

A large part of Norway's western coastline consists of steep mountains, narrow fjords and a conglomerations of islands. Fishing and farming have historically been the most important trades. Early industrial development was closely linked to primary production. Transportation along the extended coastline encouraged the establishment of a shipbuilding industry which is still important in this area, along with other industrial trades.

High mountains with rivers provided cost efficient resources for the production of electricity. This has encouraged the development of various manufacturing industries. The offshore oil activity which dominates today's Norwegian economic development has had a tremendous impact on the development of this area.

The Norwegian offshore market is dominated by a few huge national super suppliers. They are called super suppliers because they undertake the primary responsibility for very large offshore projects through new forms of contracts. One of the members of the TfS network is such a major super supplier to the oil companies. This super supplier plays an exclusive role in the industrial and economical development of the region.

Another category of more traditional supplier is situated in this area, and some of them are members of the TfS network.

The TfS network represents around 10,000 employees and a total sales volume of several billion Norwegian Kroner. Altogether there are 12 companies in the TfS network; seven of these are so-called core enterprises in the ED 2000 project at RF. They are identified as core enterprises because they have committed themselves to establishing TQM processes in close co-operation with other companies in the network organization, as well as with RF in the context of the research program. The other companies participate in a less intensive manner.

The Technological Cooperation (TESA) network

The TESA network is the oldest known formal enterprise network in Norway, dating back to 1957. It is located along the west coast of Norway and is situated to the south of the TfS network. A plateau along this coastline became one of the most productive agricultural areas in Norway. Stavanger, Norway' 'oil capital', is the main town in this area. A combination of oil and agriculture dominates the present business environment. The offshore activity, however, does not dominate the industrial profile of the TESA members, as is the case with TfS.

Over the years the content of the network collaboration has changed. Joint purchasing and supplier contracts, joint innovation projects and co-operation on information technology are examples of earlier network collaborative activities. Today many of the activities in the network are related to enterprise development promoted by the ED 2000 program.

TESA members are also involved in other activities that are not directly promoted by ED 2000. The TESA school[4] educates low and middle managers in logistics, economy and other important functions in enterprise development. TESA has started a forum were managers meet to discuss various strategic issues that the different member companies are confronted with. Both of these activities are closely related to ED 2000.

TESA members comprise enterprises within the engineering/mechanical industry. However, they do not form specific clusters within the network. On the other hand, some of the members have developed close customer–supplier relations, something that was quite rare until quite recently.

There are 13 members in the network, 12 of which have joined the ED 2000 project. Members of the network range from producers of relatively conventional mechanical products to manufacturers of high technology products, including producers of communication technology as well as flexible automation systems.

The main task of the enterprise development projects is to support further development within the following areas:

- Customer relations
- Process improvements and organizational change
- Competence development
- Further development of production processes and production philosophies appropriate to the enterprises involved
- Health and safety in the work environment (HSE).

The TESA network is about half the size of TfS, both in terms of sales and number of employees.

The SYNERGI network

SYNERGI consists of five of the largest enterprises in Norway. They are mainly related to the offshore market. Members are among Norway's largest industrial groups within the engineering field. Some of the other members include the biggest oil companies in Norway. The SYNERGI network employs a total of almost 90,000 persons and has a total sales volume of more than 100 billion NOK.

These enterprises have no specific regional location, but from an operational point of view, have their activities spread all over Norway as well as globally.

Safety management is the main focus of SYNERGI. Safety information and other health and safety data are collected in company specific databases. These data are kept anonymously and transferred to a central database were they are processed so that data and analyses specific to one company can be compared with data from all participating companies.

A basic assumption in safety management is that accidents are preventable through effective feedback control; i.e. through mechanisms by which information about accidents and near misses is utilized as a basis for actions to increase the level of safety. The prevalent use of accident and near miss databases in high risk industries, and the resources spent on recording and analyzing information on such incidents, are two manifestations of the importance attached to historical data in modern safety management. Recent advances in information technology have further strengthened the use that can be made of accident and near miss data. User friendly software and fast computers have given safety managers a vast amount of easily accessible data, and have made it possible to perform extensive statistical analyses in a matter of minutes. Monthly safety reports, causal analyses and the development of risk reduction measures are some well known examples of how empirical data are being used in the industry's day-to-day safety work. Enterprises utilizing this computerized information in the context of network relations and central databases thus have the possibility of comparing company specific safety management practice with the safety management experience of the companies in the network. These comparisons and analyses contribute to an increased awareness of the significance of company specific safety management policies.

The purpose of the SYNERGI network is to support development of safety considerations, safety management, safety control and accident prevention. Specifically the last mentioned topic is of significant importance in relation to ED 2000.

The emphasis on the safety part of HSE is due to the very high priority given to these issues in the offshore industry. Safety has been given as much attention as more traditional production related issues in other parts of industry. This is one of the reasons way safety is the core issue in the SYNERGI network collaboration.[5]

Flexibility paradoxes

The regions in which the networks operate have certain characteristics in common. Primary production, such as fishery and agriculture, is still important. On the other hand, these regions also have the highest density of industrialization in Norway, with more than 30% of the population employed in manufacturing industry. These features account for the possibility of combining employment in industry with farming and fishing activities.

Furthermore, the offshore industry dominates the region. This industry shows a high degree of fluctuation. Giant projects require a great temporary expansion of the workforce. These projects also have sensitive time schedules. Periods with high activity in which complex projects are undertaken are followed by more relaxed intervals.

A competent workforce is an obvious necessity if a company is to be effective as well as competitive. In times of great activity great demands are made on the workforce and on hired specialists. The hiring of poorly qualified workers is the last thing a company will engage in.

Many of the companies in the region are important suppliers to the offshore industry. Some of them present an image of *flexibility*. They emphasize their ability to mobilize resources to handle time and supply-sensitive activities as their most important, distinctive competence. One of the main consequences of this type of flexibility is a lack of systematic planning and organization of work. These enterprises act more or less *spontaneously* and in an *ad hoc* way in order to finish time- and resource-sensitive projects to satisfy their customers' demands. For these enterprises one important question has been whether or not to formalize planning and systematize handling of project to satisfy customer demand, and if so, whether these improvements will reduce the *spontaneous forms of flexibility* that constitutes the enterprise profile and company image.

The families of the workforce are an important buffer in this spontaneous way of meeting customers demands. They bear the brunt of the extra effort required from the almost exclusively male workforce. The structure and dynamics of this type of business in combination with the local traditions reproduce this gender specific division of labor which make these 'spontaneous' forms of flexibility possible.

Enterprise development in some of these enterprises has been linked to TQM. One improvement project in the TQM process focuses on greater control of the critical allocation of resources. Fluctuations in the market will be met with a greater internal capability to *structure* production. Old, 'spontaneous' flexibilities will be replaced by greater *new flexibility*, acquired through more *slack* in the organization. This increased slack will be created through a better overview of available resources, the introduction of systematic priorities among different projects, improved production provision, IT-based systems, improved working conditions, etc.

This new flexibility will have a number of consequences, among them:

- More flexible working conditions within more predictable working hours. When tasks are more clearly defined in advance, the employees themselves can determine their working hours.
- The business environment will change from being stressful to a more relaxed state.

These are some of the important preconditions that will contribute to a business environment with an increased responsiveness to market fluctuations and market demands. At the same time it will provide for more openness towards a more equal gender recruitment practice. More slack in the organization through increased planning requires less of a gender specific workforce (i.e. males) and demands less of the family in terms of being a buffer for the companies.

In short, the TQM process will contribute to new forms of flexibility that will:

- lead to less spontaneous flexibility and to improved working conditions (HSE). Increased predictability and control will provide more choice, more stability and more alternatives with regard to the organization of work.
- remove the strain put on the family by being less of a buffer.

A new TQM culture may increase the possibility for flexible forms of work organization and a more suitable cultural-structural business environment. Policies and practices compatible with these new forms of flexibility can be generated and spread through network collaboration. This process can be greatly facilitated by network members who are in the front line of enterprise development, such as the super suppliers and huge oil companies, for instance.

Research in networks

Action research in networks

A specific approach, based on action research principles, has been developed for two of the networks, TfS and TESA (Gustavsen, 1985, 1990; Pålshaugen, 1992a, 1992b). The main features of this design are as follows:

- An initial stage of introducing the approach and exploring its possibilities with potential participants. In the TfS network preliminary meetings were held with the network administration. A start-up conference was conducted to actively involve the various participants. In the TESA network meetings were held with the network administration and each of the different companies separately. Participatory links to the project were developed with each company.
- After the initial stage, a map was drawn up of the important characteristics of each of the participating companies. Mapping out each company in this way served as a way to benchmark the different companies in the network as well as providing a first stage in the identification of improvement projects. This survey was designed to cover both traditional benchmarking areas, like economy, market, production, strategy, customer–supplier relations, innovation, etc., as well as conditions concerning health and safety, employee participation, union relations, family–work relations, and so on. These latter, 'softer' aspects of the survey are related to the Scandinavian model of enterprise development mentioned earlier.

This designing of a survey based on a traditional benchmarking but which also included the conditions of the Scandinavian model characterized the entire approach. It led to the label 'benchmarking the Scandinavian way'.

- As a next step, the researchers presented the representatives of each company with the major issues and problems generated by the survey. This presentation was the starting point of a *participatory process*. All, or in the bigger companies a selection of representative employees, participated on an equal level in the strategic selection of improvement areas and projects. A specific consensus oriented communication method was applied to reach a joint decision on the major improvement projects to be implemented.

Final selection and ranging of areas and projects were decided on jointly for each company by union representatives and management, and were presented at a network conference.

- The network conference was used to identify overlapping company improvement areas. These areas were marked as specific network improvement areas. Furthermore, those issues that were common between networks were identified by the researchers as areas for further development.
- The companies were responsible for the organization and development of both the improvement projects and the TQM processes within their companies. In each company an employee was given the responsibility for coordinating the two.
- In the TfS network each improvement project is identified as part of a TQM process.

Process emphasis and enterprise development

In many ways the TFS and TESA networks resemble a company structure. Both networks have an ownership structure, an administration and specific work groups. The work groups are organized around sales, purchasing, personnel, production, economy, etc. reflecting the functional divisions in many companies. The SYNERGI network, on the other hand, is based on a much less structured form of collaboration. In many instances these various types of co-operation are all called networks, without paying attention to the different ways in which they make use of the network relations and organize their collaborative relationships.

An important point to be made here is the possibility of organizing development processes in network relations. In the ED 2000 project at RF certain concepts are used to encourage significant changes within enterprises as well as in the context of network relations. TQM, and to some extent BPR, are among the most important concepts used for this purpose. The traditional functional divisions are being challenged and a greater focus on more process oriented ways of working with development is encouraged.

Accordingly internal improvement projects in the enterprises have a *process orientation*. In the context of ED 2000 at RF, process orientation is conceived as a way to organize tasks horizontally, resembling a project organization (see for instance Hammer and Champy, 1993; Willoch, 1994). The difference from project organization, if one takes a process perspective, is that the latter does not have a fixed starting point nor a specific finishing date. The business aspects also play a major role in the way process orientation is emphasized in ED 2000 at RF.

A preliminary mapping of the different enterprises in the networks identified business processes and their potential for improvement. In the TfS network every improvement project is aimed at establishing and maintaining a TQM process in seven of the participating enterprises. TQM is *the* unifying label for all the business processes identified in participating companies.

Improvement projects with a process focus have been the mainstream philosophy in enterprise development for some time (Hammer and Champy, 1993; Willoch, 1994). Process based development among enterprises in a network or cutting across networks is something quite unique in the ED 2000 project at RF. Researchers in the project are trying to initiate process oriented improvement projects between different enterprises inside the TESA and TfS networks. They are also trying to establish projects embedded in a process oriented philosophy across the two networks.

The underlying philosophy has its roots both in the TQM and the BPR tradition. The process focus in the TfS and TESA networks challenges the traditional functional division of organizations. It draws attention to the limitations of a functional division of a network organization similar to the way BPR challenges the same kind of functionally divided structures inside an enterprise.

Competence building, organizational development and work environment improvement are important aspects in the development projects in TfS and TESA. Social aspects overshadow more technological aspects. The same is true in the SYNERGI network. Social aspects form the core of the different improvement projects. They are the key elements in the process focus, challenging the functional division of both hierarchical structures in enterprises and the functional divisions of network organizations like TfS and TESA.

This description gives an indication of the way research on networks is conducted in ED 2000 at Rogaland Research. The three networks demonstrate different approaches to utilizing network co-operation.

There are some common aspects in the three networks. In each of the networks enterprises compare themselves with other enterprises and are thus faced with new challenges. Some of the enterprises in the networks are customers, some are competitors, many are neither. The possibility of knowledge transfer from customers, suppliers and competitors is an important feature of networks (Porter, 1990; Reve *et al.*, 1992). Particularly in networks where market relations are of minor significance there is a good opportunity for inter-organizational learning.

Enterprise development through network co-operation is an obvious way for companies to enhance their knowledge and business practice. Comparisons with other companies can give insight into the various ways in which things are done in practice and may offer new directions for development.

Network co-operation provides an opportunity to:

- contribute to efficient utilization of common resources (sharing of consultancy resources, experience, knowledge, production, purchasing, etc.);
- create common arenas where companies can learn with and from each other.

Concluding remarks

The ED 2000 project at RF is designed especially to promote the development of enterprise networks. Employee participation and union involvement in combination with the application of international management concepts are the cornerstones of the project.

Modern enterprises, specifically big international companies, use IT, systems for planning, work organization, etc. to create well structured and controlled business practices. Smaller enterprises and suppliers, on the other hand, have positioned themselves as 'flexible specialists', able to handle unpredictable and less structured tasks (Piore and Sabel, 1984; Kvadsheim, 1996). They have created an image of themselves characterized by 'spontaneous' flexibility. Although super suppliers and international oil companies also have to adjust quickly to changing demands in the market, they use TQM as a way of creating more structured forms of flexibility.

One of the main advantages of network co-operation is the possibility for enterprises to utilize the experiences of bigger companies in their TQM practices. Older, spontaneous forms of flexibility are replaced by new forms of more systematic flexibility. Through comparison with other enterprises in the network, companies have an opportunity to learn and improve their ability to develop in an international competitive environment. In this process business systems and organizational tools are adjusted to the specific needs of small and medium sized manufacturers.

The ED 2000 project at RF is complex in terms both of content and structure. Some of the complexity can be illustrated through this listing of the main research topics in the ED 2000 project at RF:

- to promote networks as a vehicle in organizational development;
- to focus on the improvement of business practice, that is on effectiveness and efficiency, and to do so in the Scandinavian context of participation and collaboration within the organization as well as between enterprises;
- to focus, in other words, on both means and ends; on instrumental rationality as well as on substantive rationality (to use Weberian terms);
- to bring together management, unions, employees and research in arenas of discourse to work through the various aspects of such a joint enterprise and to form development coalitions;
- is to engage in a process which can only succeed if, in addition to the necessary rational and functional arrangements, there is trust, or rather the conditions for developing trust.

Trust is the invisible glue, the silent condition underlying collaboration. Shared interest, the obvious need for pooling resources, for exchanging information; all the usual reasons for network development are not sufficient if there is no trust, which is the crucial condition which enables participants to deal with the inevitable unfolding and unpredictable nature of a development process. Trust is not something one can bring or introduce, it has to be built, it is the integral component in and of the development process. This is where action research with its joint involvement and its dialogical approach derives its strategic meaning. In the problem driven orientation of action research, the researcher, while engaged in developing a relationship of partnership, is also involved in a process of developing trust, and as such is contributing significantly to the 'trust structure' underlying the development of networks, the main objective of the ED 2000 project at RF.

Notes

1. Rogaland Research is one of Norway's largest polytechnic research institutes, located in the southwestern part of the country. The primary research interests of the institute focus on the on- and offshore industry.

2. Total Quality Management (TQM), as understood in this paper, refers to a concept based on direct participation. TQM consists of some universal standards (Juran, 1964/1985) to improve quality through people (Imai, 1986). Business Process Reengineering (BPR) is used here to refer to a management concept that focuses especially on more radical changes than the more stepwise changes often associated with TQM (Hammer and Champy, 1993).

3. Each of the abbreviations refers to the name of a network. The first network is an enterprise network, originally devoted to technological cooperation in one specific region: Sunnhordland. The name of the second network, Technological Cooperation, refers to the original technological orientation in the collaborative effort among enterprises in a region located to the south of the first network. The name of the third network is associated with the synergetic advantages of a collaborative effort among large offshore and onshore enterprises related to safety management.

4. The TESA school is a joint collaborative effort between the members of the TESA network and a local business school, aimed at the education of low and middle managers.

5. For a more detailed outline of safety management research done in SYNERGI, see Ringstand and Grundt (1994).

References

Atkinson, J. 1984. *The Flexible Firm and the Shape of Jobs to Come*. Labour Market Issues No. 5. Oxford: The Trade Union Research Center. Ruskin College.

Claussen, T. 1996. *Ingeniørtradisjoner og sosiale relasjoner, myter og realiteter*. Bakgrunn for arbeidslivsforskningen. Nordisk konference om Arbejdsliv 26–29 September 1996. København Universitet, Energicenteret, Nykøbing Falster.

Demming, E.W. 1986. *Out of Crisis*. Cambridge, Mass.: Harvard University Press.

Granovetter, M. 1985. "Economic action and social structure: The problem of embeddedness". *American Journal of Sociology* 3:481–510.

Gudmundsson, G. 1993. *Den nordiske model. En afklaring av begreber*. NordFram rapport, København.

Gustavsen, B. 1985. "Workplace reform and democratic dialogue". *Economic and Industrial Democracy* 6: 461–479.

Gustavsen, B. 1990. *Strategier for utvikling av arbeidslivet*. Oslo: TANO.

Hammer, M. and Champy, J. 1993. *Reengineering the Corporation*. London: McGraw-Hill.

Imai, M. 1986. *Kaizen*. New York: McGraw-Hill.

Juran, J.M. 1964/1995. *Managerial Breakthrough: The Classic Book on Improving Management Performance*. Revised ed. London.

Juran, J.M. 1988. *Juran's Quality Control Handbook*. NY.

Kvadsheim, H. 1996. "Fleksibilitet og byråkrati i små og mellomstore bedrifter". Nordisk konference om Arbejdsliv 26–29 September 1996. København Universitet, Energicenteret, Nykøbing Falster.

Lewin, K. 1951. In D. Cartwright (ed), *Field Theory in Social Science: Selected Theoretical Papers*. NY.

Øverland, E. 1995. "Nordic globalism". Paper presented at the Pre-Conference Workshop at the annual meeting of the Academy of Management, 4–9 August, WRI, Oslo.

Pålshaugen, Ø. 1992a. *Som sagt så gjort?* Språket som virkemiddel i organisasjonsutvikling og aksjonsforskning. Oslo: NOVUS Forlag,

Pålshaugen, Ø. 1992b. "Aksjonsforskning: en nyttig vitenskap?" *Tidsskrift for samfunnsforskning* 33: 231–251.

Piore, M. and Sabel, C.F. 1984. *The Second Industrial Divide. Possibilities for Prosperity.* NY: Basic Books.

Pollert, A. 1991. *Farewell to Flexibility?* Oxford: Basil Blackwell.

Porter, M. 1990. *The Competitive Advantage of Nations.* N.Y.

Ringstad, A.J. and Grundt, H.J. 1994. "The SYNERGI project". In J. Flamm (ed), *Proceedings of the Sixth ESReDA Seminar on Maintenance and System Effectiveness.* Chamonix: European Commission Joint Research Centre, 181–192. Schiller, Venneslan, Ågotnes, Bruun, Nielsen, Tölborg. 1993. *The Future of the Nordic Model of Labour Relations.* Nord: Nordisk Ministerråd.

Senge, P.M. 1991. *Den femte disiplin: kunsten å utvikle den lærende organisasjon.* Oslo.

Taylor, F.W. 1911. *The Principles of Scientific Management.* NY: Harper.

Taylor, D. 1995. *Business Engineering with Object Technology.* NY: John Wiley.

Quale, T.U. and Øverland, E. 1996 (forthcoming). "Norway – Out of Europe?" In Szell, Nicolaou-Smokoviti, Vianello and Vilrokx (eds), *European Industrial Relations. An Introductory Course.* Rotterdam: ERASMUS.

Venneslan, K. and Ågotnes, H. 1993. "Transnationalisation and participation". In Schiller, Venneslan, Ågotnes, Bruun, Nielsen, Tölborg (eds), *The Futrure of the Nordic Model of Labour Relations.* Nord: Nordisk Ministerråd.

Williamson, O. 1987. *The Economic Institutions of Capitalism: Firms, Markets, Relational Contracting.* NY.

Willoch, B.E. 1994. *Business Process Reengineering.* Bergen: Fagbokf.

Zairi, M. and Leonard, P. 1994. *Practical Benchmarking: The Complete Guide.* London: Chapman & Hall.

Chapter 9

Concluding Remarks

Bjørn Gustavsen

Enterprise Development 2000 is, as I stated in the introductory chapter, in an evolutionary phase. The cases just presented demonstrate how a number of research groups position themselves in relation to enterprises and workplaces within the context of forming development coalitions. In spite of participating in one and the same program, the research groups do not express one single, uniform understanding of their field and what needs to be done. Rather, they embark on routes that generally differ and have relatively few direct points of contact. This is not a-typical. Research and development in working life generally is, today, a panorama of 'schools', 'positions' and 'perspectives'. 'Theory wars' are still fought, but they tend to be less prominent than in the past. It is not quite correct to say that developmentally oriented research is becoming more 'local' since much of what emerges still lays claim to some kind of universal validity — cf., for instance, the debate surrounding the concept of 'lean production' — but it certainly becomes more contextual, more imbedded in specific social realities. On a modest scale, and within the context of one single society — and a small one at that — ED 2000 reflects this type of situation.

In spite of this initial pluralism and the associated lack of contacts between the research groups, there are also elements of a joint agenda. A major part of this agenda revolves around the issue of democracy/participation in the light of the demands imposed by global competition and the associated development of new concepts of productivity, such as total quality, zero stock production and global production networks. How do the traditional Norwegian/Scandinavian patterns of labor-management co-operation stand in relation to processes of this kind: are they an asset, are they a liability, or are they irrelevant?

What can be seen from the above chapters is that this problem does not lend itself to a simple yes or no answer. Rather, we have to penetrate a number of fields, understand what happens, and create and test new ways of practicing labor–management co-operation. Without this complex procedure the main question will escape clarification.

Given problems of such a high complexity, the pluralist pattern demonstrated by research is not inadequate. Rather, this pattern makes it possible to explore a number of different aspects of the main problem and do it in such a way that each part-issue, part-theme, is approached by researchers with an interest in this theme. In Enterprise Development 2000 as it emerges above, some of the groups — such as the Norwegian School of Economics and Business Administration and the Trades Unions' Research Organization as well as the Work Research Institute — can explore the internal arenas and dynamics of co-operation. Others, such as Nordvest Forum and the Rogaland Research Foundation, can work with the network and inter-enterprise perspectives and clarify the significance of co-operation in this context. Still others — such as the Agder Research Foundation — can explore the managerial side of the issue; the Trondheim group can approach the man–technology interaction, and so on. In this way a many-faceted picture can emerge where each facet is explored on its own premisses, by a research group dedicated to this facet and to the development of views that are as adequate as possible on just their own chosen topic. This ensures an exploration which is radically different from the one emerging when one single research group has to cover everything, and from the standpoint of proving the validity of one single theoretical understanding. Even if this ambition is left aside, it will be hard for a single group to explore all aspects with an interest and enthusiasm that is equally prominent in relation to all themes and issues.

Research in working life has evolved along different trajectories. A main one can, in a somewhat simplified form, be presented as follows:

In working life a number of unfortunate practices could be observed, in the form of monotonous jobs, a low degree of communication between work roles, authoritarian forms of management, and a great deal more. These 'malpractices' were, in turn, thought to be linked to a faulty understanding of the nature of workplaces and of what productivity and good work demands. This faulty understanding could, on its part, be reduced to a theory — generally referred to as Taylorism. (Taylorism is here taken as a metaphor with no effort to penetrate its various meanings.) The problem was to create alternative practices. To achieve this, however, it was thought necessary to create an alternative *theory* that could demonstrate a superior rationality. Consequently, the social sciences embarked on the project of creating this new rationality.

In this, many succeeded. This, however, was also the problem: the fact that many succeeded led, even at an early stage, to a pluralist theoretical picture. The human relations movement, emphasizing the social relationships in the workplace, can be said to be the first 'school' of a sufficient magnitude to attack the problems of working life on a broad front. This was followed by the sociotechnical school, and, for that matter, other schools which initially based themselves on the cross-disciplinary field called general systems theory. With the advent of Bravermans ‘Labor and monopoly capital’ (Braverman, 1974) the foundation was laid for the labor process school, and so on. Other lines of theoretical development followed other trajectories, such as contingency theory, a main point of which is that there is no single organizational optimum to be identified for all kinds of enterprises. There is a need, for instance, to differentiate according to technology and market conditions. In this way, contingency theory became a general theory with internal differentiation, paving the way for such more radical forms of contingency thinking as can be found in the notion of ‘local theory’. For all, however, it was assumed that each school constituted a ‘theoretical universe’ — a universe of approaches and understandings that could be identified in terms of its own characteristics, or — to use the probably most famous word from the modern theory of science — in terms of a paradigm. The problem was, of course, that no single school could lay claim to the high ground. And what, then, about the claim to represent the universally rational? There is, of course, only one answer: all schools had to accuse the others of being mistaken. As universal rationalities go, only one school can, by definition, be the true one — the others have, in some way or another, to be faulty.

While theory battles were fought in the academic community, the processes evolving in working life seem to have been little influenced by the outcomes of these battles. The hope that Taylorism could be outwitted — or forced out — in one sweeping movement through one brilliant general theory proved to be beyond practical realization. Insofar as Taylorism was forced into retirement it was not because of the force of one single theory but rather because of the force of human experiences. As a mismatch was experienced between the principles and advice of Taylorism on the one hand and work practices such as people wanted to experience them on the other, efforts were made to reduce this mismatch. The efforts were of many different kinds: some put their faith in new forms of more humane leadership, others in more job freedom, still others in interference from society via the state in terms of stronger legislation, yet others in the ability of the unions to force new patterns, etc. All these approaches had some degree of success — they all gave something to someone, but not everything to everybody. The point, however, is another one — the efforts generally emerged out of practical experiences and took the form of mobilizing actors to change practices. The use of ‘theory’ as an intervening variable between practices and actors was very limited. Cases can be found where theory played a role, but even when cases are found — such as in the design of the Volvo Uddevalla factory (Ellegård et al., 1989) — it is seen that the work of the design group was not controlled by one single theory: several aspects of this kind were brought to bear, in an eclectic fashion, on the design of the plant.

Work life research with a practical intent can, in sum, be said to face two core problems.

First, the problem of seeking recourse to reason and rationality to legitimize analyses and proposals when there are several such competing rationalities.

Second, the problem of penetrating practices: even if one branch of research should be able to rule the roost within a certain culture at a certain point in time, it still would not be ensured an automatic penetration of actual practices.

In designing ED 2000 these problems were brought to the forefront. This was done, furthermore, by redefining them away from being problems, becoming instead what can be called natural conditions for social research in working life. As points of departure for the program it was taken for granted that no single uniform rationality to guide all workplace level processes can be developed and, second, that, on the part of the workplace actors, the change-generating engine is the experiences of those who are themselves concerned. Such experiences can be of many different kinds — often they start with production problems; things do not go as expected, there are stoppages, errors and other events that put all concerned under pressure. In other cases it can be high rates of absenteeism, injuries or other indicators of health and safety problems. In still other cases it can be a feeling of not being able to act adequately on the international scene — contracts are lost because of language problems, cultural misunderstandings, lack of ability to make decisions in tune with other decision makers and the like. Out of such experiences of mismatches, problems and needs, there emerge processes of practical reconstruction. Practical reconstruction is not primarily a process of theoretical learning that, after a lengthy learning session, brings forth a new solution in the form of an implementation of theory. Rather, the process of practical construction is *the* process — this is the main axis, or trajectory, of what is done. Theoretical understandings and related elements can be input into this process, however.

In interpreting and giving meaning to practical experiences, it turned out that research could play a role. The research that proved unable to offer one single, coherent and revolutionary plan of attack on prevailing practices turned out to be quite helpful if it could be linked to concrete, specific experiences to help give these experiences meaning. Since local work experiences, even in an interpreted form, do not primarily emerge as coherent and stable structures but rather as ongoing, fluid processes, the task of placing existing experiences within a figure–ground relationship often calls for local action. It is difficult to assign meaning and value to certain ways of doing things unless they can be compared to other ways, in terms of alternative practices. In this way interpretative local research was pulled into a role that came to resemble the more classical notion of 'action research', in the sense that research came to share responsibility for practical action as a part of its interpretative task. This role as a local interpretative resource did not carry the same glamour as a new and successful world theory, but it could provide a living while waiting for the world theory. This is also where interpretative local development research came to part from the notion of action research: as originally developed, this concept was coined to reflect the experimental demonstration of universal rationality (Gustavsen, 1996).

The potential of such a pluralist research strategy, as represented by Enterprise Development 2000, to provide premisses for a 'national agenda' remains to be demonstrated. There is a possibility, for instance, that the labor market parties will say that what can be learned from this program is so disparate, and so difficult to gain a view of, that it is not useful in our further development of, for instance, the agreement framework for labor–management co-operation. However, even if the labor market parties and other national actors end up drawing this conclusion, it is not necessarily the case that they would have been better served by a uniform theory. In such activities as redesigning the agreement system the social partners do not pursue a simple deductive logic based on identifying some simple characteristics of the global context, some simple characteristics of the co-operation system, and then drawing conclusions as if they could emerge out of some sort of syllogistic reasoning. Such processes of legal and procedural redesign are very complex ones that initially take the form of discourses. In these discourses the social partners may express different views, not to say very complex views, since the social partners are federative umbrellas, sheltering a large number of membership organizations which will hardly speak with one voice. It is in this immensely complex panorama of discourses that research will have to move, gradually playing itself into the processes. And to do this it is not certain that research should speak with only one voice, either. It may be that, on the workplace level, it is the task of research to help explore and support various arguments. All this, of course, provided that the requisite agreements can be reached in the end, so that new rules and practices can be established. And the ability to reach this agreement hangs on the ability of all discussants to accept certain rules concerning argumentation, willingness to listen to others, to consider facts and the like. Even though the moving force is a discourse rather than a universal logic of reasoning, where it is fully possible to be a successful participant even without one single view, systematic efforts will, as stated in the introduction, be made to bring the various research groups more closely together. It is an aim to create more and better interfaces and to get the research groups to resemble a quilt where all groups play a role in a larger whole. If views can emerge from this process that can gain the support of all the research groups it is, of course, a major advantage. The point is not to prevent general views from emerging, it is to let them emerge in a way that is consistent with the nature of the field to which the views pertain. Even if such views can be created, they will, however, not be able to demand the automatic acceptance of other parties. The other parties maintain their right to proceed by discourse, including the right to dispute the views of an agreed research community. For research there is little reason to fear that it will be caught up in some kind of absolutist battle between the labor market parties, where all the points and arguments of research will be torn apart. The labor market parties have a long tradition of reaching agreement just in situations where opinions differ and complexity reigns. Actually, if one is to judge by history, it is research that often needs to learn the principles of a democratic dialogue.

References

Braverman, H. 1974. *Labor and Monopoly Capital.* Monthly Review Press, New York.

Ellegård, K., Engstrøm, T. & Nilsson, L. 1989. *Principer och realiteter: Projekteringen av Volvos bilfabrik i Uddevalla.* Arbetsmiljøfonden, Stockholm.

Gustavsen, B. 1996. Development and the social sciences — an uneasy relationship. In: Toulmin, S. & Gustavsen, B. (eds): *Beyond Theory: Changing organizations through participation.* John Benjamins, Amsterdam/Philadelphia.

About the Authors

Bergur Elias Ágústsson is born in Iceland 1963. He has work experience as fisherman and in fishing industries both in Iceland and Norway. He has a master degree in Fishery Science (fiskerikandidat) from Norwegian College of Fishery Science, University of Tromso, 1992. (Thesis title: Productivity Measurement in the Fishing Industry). From 1992 to 1996 he worked at the Norwegian Institute of Fisheries and Aquaculture, and from 1996 at the University of Tromso. He has published working papers, reports and conference proceedings mainly on benchmarking.

Odd Jarl Borch is a Senior Researcher at Nordland Research Institute in Bodø, and a Research Coordinator at the Norwegian Research Council. He is a former Associate Professor of Management at Bodø College. He was a Visiting Scholar at Indiana University 1992-93. He received his M.Sc. from The Norwegian School of Economics and Business Administration in Bergen, and his PhD from Umeå University in Sweden (1991). Dr. Borch is conducting research within the field of competitive strategy, strategic alliances, and organizational development. He has published in various international journals, including Advances in Strategic Management, Journal of Management Studies, Entrepreneurship, Theory, and Practice, Scandinavian Journal of Management, and Decision Support Systems.

Erik Borgen is research scientist at Department of Production Engineering in SINTEF Technology Management. He holds a degree in information systems.

Tom Colbjørnsen born 1951, Dr. Philos. in sociology from the University of Bergen in 1984. Colbjørnsen was professor in sociology at the University of Bergen from 1981 to 1987, and is now professor in organization sciences at the Norwegian School of economics and Business Administration. His research has concentrated on issues of unemployment and labor market efficiency, and on organizational designs of firms. He has published several articles and books on these issues.

Tor Claussen is 47 years of age, a social anthropologist with a background in enterprise development and computer science. He will complete his Ph.D in philosophy in 1999. He works as a senior researcher at Rogaland Research in

Stavanger, Norway. He is currently a project leader of one of the biggest social science based projects in enterprise development in Norway, involving more than 12 researchers and 30 enterprises with approximately 100,000 employees.

Eivind Falkum born 1952. Cand. polit. at the University of Bergen, Institute of Administration and Organization Science 1980. Lecturer same place till 1984. Then research scientist in the SINTEF-group for ten years, first occupied with studies, design and evaluation of programmes in the field of the welfare state, later on studies on the impact of information technology on management and organisation. From 1994 research scientist at Fafo (Institute of Applied Social Sciences) occupied with industrial relation studies and leader of the Fafo part of "Enterprise Development 2000".

Lene Foss is a Senior Researcher at the Norwegian Institute of Fisheries and Aquaculture Ltd, Centre of Economics and Marketing in Tromso. She holds an MBA (1989) and a Ph. d (1994) from the Norwegian School of Economics and Business Administration in Bergen. Her research area is entrepreneurship, social networks and organizational design. She has published articles on entrepreneurship and aquaculture. Foss is currently a Visiting Scholar in the Department of Sociology,University of North Carolina at Chapel Hill, USA.

Reidar Gjersvik is research scientist at the Departnment for industrial Economics and Technology Management at Norwegian University of Science and Technology. He holds a Ph.D. in organization.

Bjørn Gustavsen Dr. Philos., is Professor at the National Institute for Working Life in Sweden and Research Advisor to the "Enterprise Development 2000" program under the Norwegian Reseach Council. His main field is the organization of tripartite work life development programs and associated reseach questions, such as the relationship between theory and practice, the dynamics of change, and the possibility of enlightenment and progress.

Jon Hanssen-Bauer Ph. D., is Research Director for the international department at the Fafo Institute for Applied Social Science in Oslo, Norway.
As a research fellow at the Work Research Institute in Oslo from 1983 to 1989, he researched and published on such topics as organizational design, work environment, socio-technical optimization, and management of project organizations for design of offshore oil and gas installations. From 1989 to 1993 he worked for the Norwegian Work Life Center (SBA), an enterprise development program supported by the Norwegian labor market partners and the Government.

Roger Klev is research scientist at SINTEF-IFIM. He holds a Ph.D. in organization.

Harald Knudsen born 1943, is professor of management at Agder College, Kristiansand. He is educated in business administration, and has an MBA and a Ph.D. degree from the University of Oregon. Since 1976 he has worked at Agder College. His present teaching and research interests are in the areas of strategic and international management, leadership development and organizational learning.

Morten Levin is professor at the Department of Sociology and Political Science at the Norwegian University of Science and Technology in Trondheim. He holds a master degree in engineering and a graduate degree in sociology.

Ida Munkeby is research scientist at SINTEF-IFIM. She holds a degree in sociology.

Øyvind Pålshaugen 1951, Dr. philos. Researcher at the Work Research Institute, Oslo. His field of study includes action research, organization development, the rhetorics of science and the science of rhetoric. He is co-editor of the journal "Concepts and Transformation".

Monica Rolfsen is research scientist at SINTEF-IFIM. She holds a Ph.D. in organization.

Knut Senneseth Master of Science in Business, works as a researcher at Nordland Research Institute, where he mainly works in the field of entrepreneurship and businessdevelopment.

Hans Jørgen Sæbø is research scientist at SINTEF-Safety and Reliability. He holds a degree in engineering.

In the series DIALOGUES ON WORK AND INNOVATION the following titles have been published thus far or are scheduled for publication:

1. NASCHOLD, Frieder and Casten VON OTTER: *Public Sector Transformation: Rethinking Markets and Hierarchies in Government.* 1996.
2. TOULMIN, Stephen and Björn GUSTAVSEN (eds): *Beyond Theory. Changing organizations through participation.* 1996.
3. GUSTAVSEN, Björn, Bernd HOFMAIER, Marianne EKMAN PHILIPS and Anders WIKMAN: *Concept-Driven Development and the Organization of the Process of Change. An evaluation of the Swedish Working Life Fund.* 1996.
4. MERRELYN, Emery: *Searching. The theory and practice of making cultural change.* n.y.p.
5. PÅLSHAUGEN, Øyvind, Björn GUSTAVSEN, Dag ØSTERBERG and John SHOTTER: *The End of Organization Theory?* n.y.p.
6. GUSTAVSEN, Björn, Tom COLBJØRNSEN and Øyvind PÅLSHAUGEN (eds): *Development Coalitions in Working Life. The 'Enterprise Development 2000' Program in Norway.* 1998.